SpringerBriefs in Electrical and Computer Engineering

T0238873

For further volumes:
http://www.springer.com/series/10059

Tariq Jamil

Complex Binary Number System

Algorithms and Circuits

Springer

Tariq Jamil
Department of Electrical
  and Computer Engineering
Sultan Qaboos University
Muscat
Oman

ISSN 2191-8112   ISSN 2191-8120   (electronic)
ISBN 978-81-322-0853-2   ISBN 978-81-322-0854-9   (eBook)
DOI 10.1007/978-81-322-0854-9
Springer New Delhi Heidelberg New York Dordrecht London

Library of Congress Control Number: 2012949074

Printed on acid-free paper

Springer is part of Springer Science+Business Media (www.springer.com)

This work is dedicated to
my parents
my family
and
my teachers

Preface

This work is a synopsis of research work done by me and my fellow co-investigators in the fields of computer arithmetic and computer architecture spanning a period of over 20 years.

During the 1990s, discussion among the computer architects used to be focused on weighing the merits and demerits of control-flow and data-flow models of computation for parallel processing. As a doctoral student of computer engineering at the Florida Institute of Technology (USA) at that time, I became interested in devising a better model of computation which would amalgamate the best features of data-flow model with the content-addressability features of the associative memories. These efforts resulted in formulating the concept of *associative dataflow* and, consequently, the design and implementation of an *associative dataflow processor* by me in 1996.

In 1999, while at the University of Tasmania (Australia), Neville Holmes, a colleague of mine in the School of Computing, showed me a paper written by Donald Knuth, a pioneer in the field of computing, published in the *Communications of the ACM* advocating a binary number system with a base other than 2. This kindled my interest in computer arithmetic and I started doing further research in this avenue of computing. During this investigation, I found out about Walter Penny's proposal for a $(-1 + j)$ base number system which appeared more promising to me and Neville than Donald Knuth's idea. We called $(-1 + j)$ base number system as the *Complex Binary Number System* (*CBNS*) and what followed in the next 12 years of my work on CBNS is now in your hands.

During the past several years, I have worked as principal investigator on several research grants provided by Sultan Qaboos University (Oman) in an effort to establish CBNS as a viable number system. This has resulted in the publication of several conference and journal papers authored by me and my co-investigators and, in this book, I have tried to compile a succinct summary of all these publications for the benefit of anyone interested in continuing research in this area of computer arithmetic. An innovative patent on *complex binary associative dataflow processor* has been granted to me by the Australian Patent Office in 2010 which incorporates CBNS within the associative dataflow processor designed by me earlier.

It is sincerely hoped that this book will give new impetus to research in computer arithmetic and parallel processing and will enable the researchers of tomorrow to improve and implement CBNS within the realm of computing.

Muscat, Oman, 7 July 2012   Tariq Jamil

Acknowledgments

First of all, I am eternally grateful to Almighty Allah for His countless blessings in completing this task. Without the sacrifices and selflessness of my parents, it would not have been possible for me to be able to get an engineering education, and without the guidance from my teachers, I would not have been able to excel in my search for knowledge. My wife has been a continuous beacon of encouragement to me while my children, Hamiz and Umnia, are the main sources of happiness for me in this life.

I am grateful to two of my high-school mathematics teachers, Mr. Tariq and Mr. Baig, for kindling my interest in mathematics during those years of my life. My Ph.D. supervisor, Dr. Deshmukh, helped me tremendously during my research work on the topic of *associative dataflow* at the Florida Tech (USA) during 1994–1996.

I am thankful to my friend and colleague, Neville Holmes, for introducing me to the realm of computer arithmetic and also for collaborating with me in some of the earlier publications on this topic. David Blest was among the first mathematicians who saw potential of further research in CBNS and wrote papers with me in a few publications.

Australian Research Council was the first agency to support my research on associative dataflow processing during 1997–1998 and I am thankful to them for their financial support. I am grateful to Sultan Qaboos University (Oman) for supporting my research activities on the topic of CBNS through various internal research grants during the period 2000–2012. Dr. Bassel Arafeh, Dr. Amer AlHabsi, Dr. Amir Arshad Abdulghani, Dr. Usman Ali, Mr. Ahmad AlMaashari, Mr. Said AlAbri, and Ms. Sadaf Saeed have worked as co-investigators with me on these research projects and to these researchers, I express my unbridled gratitude for their work. It is due to the efforts of these people that CBNS has matured to become a viable alternative to traditional binary number system.

Finally, I would like to thank Springer for giving me the opportunity to publish this work for the benefit of the computer scientists and engineers worldwide.

Muscat, Oman, 7 July 2012   Tariq Jamil

Contents

Chapter 1
Introduction

Abstract Complex numbers play a truly unique and important role in the realm of modern science and engineering. Using these numbers it is possible to locate a point within two dimensions of a Cartesian co-ordinate system. Therefore, these numbers are used extensively in digital signal processing algorithms and image processing applications. In this chapter we are going to review basic theory about complex numbers and arithmetic operations involving such type of numbers. This will enable us to justify the need for a more efficient representation of these numbers in computer architecture.

1.1 What is a Complex Number?

A complex number represents a point in a two-dimensional system and, in rectangular form, is written in the form $(x + jy)$ where x, called the real part, denotes the position along the horizontal axis and y, called the imaginary part, denotes the position along the vertical axis. "j" is considered equivalent to $\sqrt{-1}$ and is used to represent the imaginary nature of the variable y. Another way of locating the point in a two-dimensional system is to use polar notation $r\angle\theta$ where r represents the hypotenuse of length $\sqrt{x^2 + y^2}$ for a right-angled triangle having x as the base and y as the perpendicular, and θ represents the angle made by the hypotenuse with the base of the triangle, given by $\tan^{-1}\frac{y}{x}$.

T. Jamil, *Complex Binary Number System*,
SpringerBriefs in Electrical and Computer Engineering,
DOI: 10.1007/978-81-322-0854-9_1, © The Author(s) 2013

1.2 Arithmetic Operations Involving Complex Numbers

Arithmetic operations involving two complex numbers $(a + jb)$ and $(c + jd)$ are carried out as follows:

- Addition involves two individual additions, one for the real parts $(a + c)$ and one for the imaginary parts $(b + d)$:

$$(a + jb) + (c + jd) = (a + c) + j(b + d) \qquad (1.1)$$

- Subtraction involves two individual subtractions, one for the real parts $(a - c)$ and one for the imaginary parts $(b - d)$:

$$(a + jb) - (c + jd) = (a - c) + j(b - d) \qquad (1.2)$$

- Multiplication involves four individual multiplications ac, ad, bc, bd, one subtraction $ac - bd$, and one addition $ad + bc$:

$$(a + jb) \times (c + jd) = ac + j(ad + bc) + j^2 bd = (ac - bd) + j(ad + bc) \quad (1.3)$$

- Division involves six individual multiplications ac, ad, bc, bd, c^2, d^2, two additions $ac + bd$ and $c^2 + d^2$, one subtraction $bc - ad$, and then two individual divisions $\frac{ac+bd}{c^2+d^2}$ and $\frac{bc-ad}{c^2+d^2}$:

$$\frac{(a + jb)}{(c + jd)} = \frac{(a + jb)}{(c + jd)} \times \frac{(c - jd)}{(c - jd)} = \frac{ac + j(bc - ad) - j^2 bd}{c^2 - j^2 d^2}$$

$$= \frac{(ac + bd) + j(bc - ad)}{c^2 + d^2} = \frac{ac + bd}{c^2 + d^2} + j \frac{bc - ad}{c^2 + d^2} \qquad (1.4)$$

1.3 Justification for Complex Binary Number System

Let's assume that each individual addition/subtraction, involving complex numbers, takes p ns to complete and each individual multiplication/division takes q ns to execute, such that $p \ll q$ (multiplication can be assumed to be *repeated-addition* and division can be assumed to be *repeated subtraction*), then each complex addition/subtraction will take $2p$ ns, each complex multiplication will take $4q + p + p = (2p + 4q)$ ns, and each complex division will take $6q + 2p + p + 2q = (3p + 8q)$ ns. Now imagine a number system in which complex arithmetic does not involve any *combination of individual arithmetic operations* as described in Sect. 1.2. That is, addition, subtraction, multiplication, or division of complex numbers is just one *pure* addition, one *pure* subtraction, one *pure* multiplication, or one *pure* division operation respectively and not a combination of various individual operations within a given arithmetic operation as mentioned previously. This will effectively reduce the complex addition/subtraction time to p ns and complex multiplication/division time to q ns. Mathematically, such a complex number system will yield reduction in execution time of addition/subtraction operation roughly by a factor of $\frac{p}{2p} \times 100 = 50\,\%$, for multiplication $\frac{(2p+4q)}{q} \times 100 = \frac{4q}{q} \times 100 = 400\,\%$ (since p is very small compared to q), and for division $\frac{(3p+8q)}{q} \times 100 = \frac{8q}{q} \times 100 = 800\,\%$. With the reduction in execution times of complex arithmetic operations roughly by factors of 50–800 % in digital signal and image processing applications, it is possible to achieve tremendous enhancement in the overall performance of systems based on these applications, provided a technique exists which treats a complex number as a *single entity* (rather than *two entities* comprising of real and imaginary parts) and facilitates a single-unit representation of complex numbers in binary format within a micro-processor environment (rather than *two individual representations* for real and imaginary parts respectively, as in today's computers). Such a unique number system is referred to as *Complex Binary Number System* (CBNS).

1.4 What is Complex Binary Number System?

Efforts in defining a binary number system (0 or 1) with bases other than 2, which would facilitate a single-unit representation of complex numbers, date back to 1960 when Donald E. Knuth described a "quater-imaginary" number system with base $2j$ and analyzed the arithmetic operations of numbers based on this imaginary base [1]. However, he was unsuccessful in providing a division algorithm and considered it as a main obstacle towards hardware implementation of any imaginary-base number system.

Walter Penney, in 1964, attempted to define a complex number system, first by using a negative base of -4 [2], and then by using a complex number $(-1 + j)$ as the base [3]. However, the main problem encountered with using these bases was again the inability to formulate an efficient division process. Stepanenko, in 1996, utilized the base $j\sqrt{2}$ to generate real parts of complex numbers by taking even powers of the base and imaginary parts of complex numbers by taking odd powers of the base [4]. Although partly successful in resolving the division problem as an "all-in-one" operation, in his algorithm "…everything…reduces to good choice of an initial approximation" in a Newton–Raphson iteration which may or may not converge.

Jamil et al., in 2000, revisited Penney's number system with base $(-1 + j)$ and presented a detailed analysis of this number system, now called Complex Binary Number System (CBNS) [5]. During the past several years, Jamil et al. have obtained research grants and published several articles in international conferences and journals describing conversion algorithms, arithmetic operations, and computer hardware circuits involving CBNS. This book is a compilation of the entire research work carried out on CBNS by the author and his various research teams, and is intended to assist new researcher in the fields of computer arithmetic and computer architecture as well as any student with interest in digital logic towards advancing modern day computing through incorporation of CBNS in both the software and hardware paradigms.

In Chap. 2, algorithms to convert a given complex number into CBNS are presented. This is followed by presentation of techniques in Chap. 3 to carry out the arithmetic and shift operations in the new number system. Chapter 4 describes the hardware implementation, and performance statistics related to arithmetic circuits and, in Chap. 5, incorporation of these circuits within an associative dataflow environment to design a Complex Binary Associative Dataflow Processor (CBADP) has been explained. Conclusion and further research are outlined in Chap. 6.

References

1. D.E. Knuth, An imaginary number system. Commun. ACM **3**, 345–347 (1960)
2. W. Penney, A numeral system with a negative base. Math. Student J. **11**(4), 1–2 (1964)
3. W. Penney, A binary system for complex numbers. J. ACM **12**(2), 247–248 (1964)
4. V.N. Stepanenko, Computer arithmetic of complex numbers. Cybernet. Syst. Anal. **32**(4), 585–591 (1996)
5. T. Jamil, N. Holmes, D. Blest. Towards implementation of a binary number system for complex numbers. in *Proceedings of the IEEE Southeastcon*, 268–274 (2000)

Chapter 2
Conversion Algorithms

Abstract In this chapter, algorithms for conversion of complex numbers into complex binary number system (CBNS) will be described. We'll start with integers, then explain how fractional numbers can be converted into CBNS, and finally how to represent floating point numbers into the new number system. Along the way, we'll also describe how imaginary numbers can be converted into CBNS. Once the algorithms for conversion of real and imaginary parts of a complex number (whether integer, fraction, or floating point) are known, we'll describe how a given complex number can be represented as single-unit binary string consisting of 0 and 1s.

2.1 Conversion Algorithms for Integers

Let's first begin with the case of a positive integer N (in decimal number system) [1, 2]. To represent N in CBNS, we follow these steps:

(i) Express N in terms of powers of 4 using the repeated division process. That is, repeatedly divide N by 4 keeping track of the remainders.
Examples:
$2012_{10} = (1,3,3,1,3,0)_{\text{Base } 4}$
$2000_{10} = (1,3,3,1,0,0)_{\text{Base } 4}$
$60_{10} = (3,3,0)_{\text{Base } 4}$
(ii) Now convert the Base 4 number $(\ldots n_5, n_4, n_3, n_2, n_1, n_0,)$ to Base -4 by replacing each digit in the odd location $(n_1, n_3, n_5, \ldots)$ with its negative to get $(\ldots - n_5, n_4, - n_3, n_2, - n_1, n_0,)$.
Examples:
$2012_{10} = (-1, 3, -3, 1, -3, 0)_{\text{Base } -4}$

T. Jamil, *Complex Binary Number System*,
SpringerBriefs in Electrical and Computer Engineering,
DOI: 10.1007/978-81-322-0854-9_2, © The Author(s) 2013

Table 2.1 Equivalence between normalized base -4 and $(-1+j)$-base CBNS representation [1]

Normalized base -4	CBNS representation base $(-1+j)$
0	0000
1	0001
2	1100
3	1101

$$2000_{10} = (-1,3,-3,1,0,0)_{\text{Base}-4}$$
$$60_{10} = (3,-3,0)_{\text{Base}-4}$$

(iii) Next, we normalize the new number, i.e., get each digit in the range 0–3, by repeatedly adding 4 to the negative digits and adding a 1 to the digit on its left. This operation will get rid of negative numbers but may create some digits with a value of 4 after the addition of a 1. To normalize this, we replace 4 by a 0 and subtract a 1 from the digit on its left. Of course, this subtraction might once again introduce negative digits which will be normalized by the previous method but this process will definitely terminate. What is interesting to note is that, with negative bases, all integers (positive or negative) have a unique positive representation.

Examples:

$$2012_{10} = (-1,3,-3,1,-3,0)_{\text{Base}-4}$$
$$= (1,3,4,1,2,1,0) = (1,2,0,1,2,1,0)_{\text{Normalized}}$$
$$2000_{10} = (-1,3,-3,1,0,0)_{\text{Base}-4}$$
$$= (1,3,4,1,1,0,0) = (1,2,0,1,1,0,0)_{\text{Normalized}}$$
$$60_{10} = (3,-3,0)_{\text{Base}-4}$$
$$= (4,1,0) = (-1,0,1,0) = (1,3,0,1,0)_{\text{Normalized}}$$

(iv) Lastly, we replace each digit in the normalized representation by its equivalent binary representation in CBNS, as per Table 2.1.

These equivalences can be verified to be correct by calculating the power series for each CBNS representation as follows:

$$0000 = 0 \times (-1+j)^3 + 0 \times (-1+j)^2 + 0 \times (-1+j)^1 + 0 \times (-1+j)^0$$
$$0001 = 0 \times (-1+j)^3 + 0 \times (-1+j)^2 + 0 \times (-1+j)^1 + 1 \times (-1+j)^0$$
$$1100 = 1 \times (-1+j)^3 + 1 \times (-1+j)^2 + 0 \times (-1+j)^1 + 0 \times (-1+j)^0$$
$$1101 = 1 \times (-1+j)^3 + 1 \times (-1+j)^2 + 0 \times (-1+j)^1 + 1 \times (-1+j)^0$$

Examples:

$$2012_{10} = (1,2,0,1,2,1,0)_{\text{Normalized}}$$
$$= 0001\ 1100\ 0000\ 0001\ 1100\ 0001\ 0000$$
$$= 1110000000001110000010000_{\text{Base}\,(-1+j)}$$

$$2000_{10} = (1, 2, 0, 1, 1, 0, 0)_{\text{Normalized}}$$
$$= 0001\ 1100\ 0000\ 0001\ 0001\ 0000\ 0000$$
$$= 11100000000001000100000000_{\text{Base }(-1+j)}$$

$$60_{10} = (1, 3, 0, 1, 0)_{\text{Normalized}}$$
$$= 0001\ 1101\ 0000\ 0001\ 0000$$
$$= 11101000000010000_{\text{Base }(-1+j)}$$

To convert a negative integer into CBNS format, we simply multiply the representation of the corresponding positive integer with 11101 (equivalent to $(-1)_{\text{Base }(-1+j)}$) according to the multiplication algorithm given in Chap. 3. Thus,

$$-2012_{10} = 11100000000001110000010000 \times 11101$$
$$= 110000000000011011101010000_{\text{Base }(-1+j)}$$

$$-2000_{10} = 11100000000001000100000000 \times 11101$$
$$= 11000000000001101000000000_{\text{Base }(-1+j)}$$

$$-60_{10} = 11101000000010000 \times 11101$$
$$= 1000111010000_{\text{Base }(-1+j)}$$

To obtain binary representation of a positive or negative imaginary number in CBNS, we multiply the corresponding CBNS representation of positive or negative integer with 11 (equivalent to $(+j)_{\text{Base }(-1+j)}$) or 111 (equivalent to $(-j)_{\text{Base }(-1+j)}$) according to the multiplication algorithm given in Chap. 3. Thus,

$$+j2012_{10} = 11100000000001110000010000 \times 11$$
$$= 100000000000010000110000_{\text{Base }(-1+j)}$$

$$-j2012_{10} = 11100000000001110000010000 \times 111$$
$$= 1110100000001110100011110000_{\text{Base }(-1+j)}$$

$$+j2000_{10} = 11100000000001000100000000 \times 11$$
$$= 10000000011001100000000_{\text{Base }(-1+j)}$$

$$-j2000_{10} = 11100000000001000100000000 \times 111$$
$$= 11101000000011101111000000000_{\text{Base }(-1+j)}$$

$$+j60_{10} = 11101000000010000 \times 11$$
$$= 111000000110000_{\text{Base }(-1+j)}$$

$$-j60_{10} = 11101000000010000 \times 111$$
$$= 11000001110000_{\text{Base} (-1+j)}$$

Having obtained CBNS representations for all types of integers (real and imaginary), it is now possible for us to represent an integer complex number (both real and imaginary parts of the complex number are integers) simply by adding the real and imaginary CBNS representations according to the addition algorithm given in Chap. 3. Thus,

$$2012_{10} + j2012_{10} = 11100000000001110000010000_{\text{Base} (-1+j)}$$
$$+ 100000000000010000110000_{\text{Base} (-1+j)}$$
$$= 11101000000001110100011100000_{\text{Base} (-1+j)}$$

$$-60_{10} - j2000_{10} = 1000111010000_{\text{Base} (-1+j)}$$
$$+ 11101000000001110111000000000_{\text{Base} (-1+j)}$$
$$= 11101000000000001101011010000_{\text{Base} (-1+j)}$$

2.2 Conversion Algorithms for Fractional Numbers

The procedure for finding the binary equivalent in base $(-1 + j)$ for real fraction and imaginary fraction is very similar to the procedure explained for integers in Sect. 2.1. As an example, CBNS representation for 0.351_{10} is obtained as follows [2]:

(i) Repeated multiplication by 4 gives:
   $0.351 \times 4 = 1.404;$
   $0.404 \times 4 = 1.616;$
   $0.616 \times 4 = 2.464;$
   $0.464 \times 4 = 1.856;$
   $0.856 \times 4 = 3.424;$
   $0.424 \times 4 = 1.696;$
   $0.696 \times 4 = 2.784;$
   $0.784 \times 4 = 3.136;$
   and so on. Thus
   $0.351_{10} = 0.11213123\ldots_{\text{Base } 4}$
(ii) Converting the Base 4 number $(\ldots n_5, n_4, n_3, n_2, n_1, n_0,)$ to Base -4 by replacing each digit in the odd location $(n_1, n_3, n_5, \ldots)$ with its negative to get $(\ldots - n_5, n_4, -n_3, n_2, -n_1, n_0,)$ yields:
   $0.351_{10} = 0.(-1)1(-2)1(-3)1(-2)3\ldots_{\text{Base } -4}$
(iii) After normalization, we have:
   $0.351_{10} = 1.32221223\ldots_{\text{Base } -4}$

Table 2.2 Equivalence between fractional coefficients and $(-1+j)$-base CBNS representation [1]

i	2^{-i}	CBNS representation base $(-1+j)$
1	2^{-1}	1.11
2	2^{-2}	1.1101
3	2^{-3}	0.000011
4	2^{-4}	0.00000001

(iv) And, finally replacing each Base -4 digit with its equivalent four-bit binary sequence as given in Table 2.1 gives:

$$0.351_{10} = 1.110111001100110000001\ldots_{\text{Base } (-1+j)}$$

Similarly,

$$j0.351_{10}$$
$$= (1.110111001100110000001\ldots) \times (11)$$
$$= 0.0100010000110100001100\ldots_{\text{Base } (-1+j)}$$

A more elaborate mathematical algorithm which can be easily programmed in some high-level language for converting fractional numbers into CBNS is described in [1]. According to this algorithm, any fraction F can be expressed uniquely in terms of powers of $1/2 = 2^{-1}$ such that

$$F = r_0 = f_1.2^{-1} + f_2.2^{-2} + f_3.2^{-3} + f_4.2^{-4} + \cdots \qquad (2.1)$$

Then the coefficients f_i and remainders r_i are given as follows:

Initially,
if $(2r_0 - 1) < 0$ then $f_1 = 0$ and set $r_1 = 2r_0$
or if $(2r_0 - 1) \geq 0$ then $f_1 = 1$
and set $r_1 = (2r_0 - 1)$
Then,
if $(2r_i - 1) < 0$ then $f_{i+1} = 0$ and set $r_{i+1} = 2r_i$
or if $(2r_0 - 1) \geq 0$ then $f_{i+1} = 1$
and set $r_{i+1} = (2r_i - 1)$

This process continues until $r_i = 0$ or the machine limit has been reached. Then, $\forall f_i = 1$, replace its associated 2^{-i} according to Table 2.2 (only the first four values of i are listed in the table; for $i > 4$, refer to Table 2.3).
As an example, let

$$F = r_0 = 0.4375_{10}$$

Initially,

$$(2r_0 - 1) = 2(0.4375 - 1) = -0.125 < 0$$

$\Longrightarrow f_1 = 0$ and $r_1 = 2r_0 = 2(0.4375) = 0.875$

Then,

$$(2r_1 - 1) = 2(0.875) - 1 = 0.75 > 0$$

$\Longrightarrow f_2 = 1$ and $r_2 = (2r_1 - 1) = 2(0.875) - 1 = 0.75$

Continuing according to the algorithm, we have

$$(2r_2 - 1) = 2(0.75) - 1 = 0.5 > 0$$

$\Longrightarrow f_3 = 1$ and $r_3 = (2r_2 - 1) = 2(0.75) - 1 = 0.5$

$$(2r_3 - 1) = 2(0.5) - 1 = 0(\text{STOP})$$

$\Longrightarrow f_4 = 1$ and $r_4 = 0$

Thus,

$$
\begin{aligned}
0.4375_{10} &= 0.2^{-1} + 1.2^{-2} + 1.2^{-3} + 1.2^{-4} \\
&= 0.(1.11) + 1(1.1101) + 1(0.000011) \\
&+ 1(0.00000001) = 1.11011101_{\text{Base}\ (-1+j)}
\end{aligned}
$$

(the addition is according to the algorithm given in Chap. 3)

It is likely that most fractions will not terminate as this example, until the machine limit has been reached, e.g.

$$0.351_{10} = 1.110111001100110000011\ldots_{\text{Base}\ (-1+j)}$$

In that case, it is up to the user to terminate the algorithm when certain degree of accuracy has been achieved.

In general, to find CBNS representation of any 2^{-i}, express i as $4s + t$ where s is an integer and $0 \le t < 4$. Then, depending upon value of t, 2^{-i} can be expressed as given in Table 2.3. All rules for obtaining negative integer and positive/negative imaginary number representations in CBNS, as discussed previously, are equally applicable for obtaining negative fractional and positive/negative imaginary fractional representations in the new base.

A complex number which has only fractional real and imaginary parts can be represented in CBNS simply by adding the CBNS representations of each part according to the addition algorithm described in Chap. 3. Thus,

$(0.351 + j0.351)_{\text{Base}\ 10}$

$= 1.110111001100110000011111001100\ldots_{\text{Base}\ (-1+j)}$

$+ 0.0100010000110100001100 1101\ldots_{\text{Base}\ (-1+j)}$

$= 0.0110100011110101111110001001\ldots_{\text{Base}\ (-1+j)}$

Table 2.3 Equivalence between value of "*t*" and $(-1+j)$-base CBNS representation [1]

t	CBNS representation base $(-1+j)$
0	$0.0\ldots(8s-1)$ zeroes followed by 1
1	$0.0\ldots(8s-1)$ zeroes followed by 111
2	$0.0\ldots(8s-1)$ zeroes followed by 11101
3	$0.0\ldots(8s-1)$ zeroes followed by 11

2.3 Conversion Algorithms for Floating-Point Numbers

To represent a floating-point positive number in CBNS, we add the corresponding integer and fractional representations according to the addition algorithm described in Chap. 3. Once again, all rules for obtaining negative integer and positive/negative imaginary number representations, as discussed previously, are equally applicable for obtaining negative floating-point and positive/negative imaginary floating-point representations in CBNS. For example,

$$60.4375_{10} = 11101000000010000_{\text{Base }(-1+j)} + 1.11011101_{\text{Base }(-1+j)}$$

$$= 11101000000010001.11011101_{\text{Base }(-1+j)}$$

$$j60.4375_{10} = \left(11101000000010001.11011101_{\text{Base }(-1+j)}\right) \times (11)$$

$$= 11100000011 0000.01000111_{\text{Base }(-1+j)}$$

Adding these two CBNS representation using the addition algorithm outlined in Chap. 3 gives:

$$(60.4375 + j60.4375)_{10}$$

$$= 11101000000010001.11011101_{\text{Base }(-1+j)}$$

$$+ 11100000011 0000.01000111_{\text{Base }(-1+j)}$$

$$= 10000011101110.1000011_{\text{Base }(-1+j)}$$

In the above example, we have been able to represent a complex number (both real and imaginary parts are floating point numbers) in a single binary string. Thus, following the procedures outlined in this chapter, we can represent any complex number into CBNS format which is characterized by a single-unit string of bits (0 or 1).

References

1. T. Jamil, N. Holmes, D. Blest, *Towards implementation of a binary number system for complex numbers. Proceedings of the IEEE Southeastcon*, pp. 268–274 (2000)
2. T. Jamil, The complex binary number system: Basic arithmetic made simple. IEEE Potentials **20**(5), 39–41 (2002)

Chapter 3
Arithmetic Algorithms

Abstract In this chapter, algorithms for performing arithmetic operations on complex binary numbers, as well as effects of shift operations on this type of numbers, will be described. We will discuss addition, subtraction, multiplication, and division algorithms for complex binary numbers along with some examples. Then, we will describe the results of implementing multiple-bits shift-left and shift-right operations on various types of complex numbers represented in CBNS.

3.1 Addition Algorithm for Complex Binary Numbers

The binary addition of complex binary numbers follows the truth table given in Table 3.1 [1–3].

Table 3.1 is similar to the half-adder truth table of traditional Base 2 binary numbers in the sense that $0 + 0$ is still 0 (represented by 0000 in four bits), $0 + 1$ is 1 (represented by 0001 in four bits), $1 + 0$ is 1 (represented by 0001 in four bits), and $1 + 1 = 2_{10}$ (represented by $1100_{Base\ (-1+j)}$ in four bits). The last case can be interpreted as follows.

When two 1s are added, the sum is 0 and (instead of just one carry as in traditional Base 2 binary addition) two carries are generated which propagate towards the two adjoining positions after skipping the immediate neighbor of the sum column.

That is, if two numbers with 1s in position n are added, this will result in 1s in positions $n + 3$ and $n + 2$ and 0s in positions $n + 1$ and n. Similar to the ordinary computer rule where $1 + 111 \ldots$ (up to machine limit) $= 0$, we have $11 + 111 = 0$, called *zero rule*, for complex binary numbers.

As an example, let us add $(1 + j)$ with $(2 - j2)$ in CBNS:

T. Jamil, *Complex Binary Number System*,
SpringerBriefs in Electrical and Computer Engineering,
DOI: 10.1007/978-81-322-0854-9_3, © The Author(s) 2013

Table 3.1 Truth table for addition of complex binary numbers

Inputs		Outputs	
Augend	Addend	Carries	Sum
0	0	000	0
0	1	000	1
1	0	000	1
1	1	110	0

$(1 + j)$ is equivalent to $(1110)_{\text{Base } (-1+j)}$ and $(2 - j2)$ is equivalent to $(111000)_{\text{Base } (-1+j)}$. Thus

$$(1 + j) + (2 - j2)$$

$$= 1110_{\text{Base } (-1+j)} + 111000_{\text{Base } (-1+j)}$$

$$= 111010110_{\text{Base } (-1+j)} = 3 - j$$

The result can be verified to be correct by calculating the power series of the complex binary number as follows:

$$1 \times (-1+j)^8 + 1 \times (-1+j)^7 + 1 \times (-1+j)^6$$

$$+ 0 \times (-1+j)^5 + 1 \times (-1+j)^4 + 0 \times (-1+j)^3$$

$$+ 1 \times (-1+j)^2 + 1 \times (-1+j)^1 + 0 \times (-1+j)^0$$

$$= 3 - j$$

3.2 Subtraction Algorithm for Complex Binary Numbers

The binary subtraction of complex binary numbers follows the truth table given in Table 3.2 [1–3].

Table 3.2 is similar to the half-subtractor truth table of traditional Base 2 binary numbers in the sense that $0 - 0$ is still 0 (represented by 0000 in four bits), $0 - 1$ is -1 (represented by $11101_{\text{Base } (-1+j)}$ in five bits), $1 - 0$ is 1 (represented by 0001 in four bits), and $1 - 1 = 0$ (represented by 0000 in four bits). The second case of $0 - 1$ can be interpreted as a special case and solved by applying the following algorithm:

Assume our minuend is:

$$a_n a_{n-1} a_{n-2} \ldots a_{k+4} a_{k+3} a_{k+2} a_{k+1} a_k 0 a_{k-1} \ldots a_3 a_2 a_1 a_0$$

and subtrahend is:

$$b_n b_{n-1} b_{n-2} \ldots b_{k+4} b_{k+3} b_{k+2} b_{k+1} 1 b_{k-1} \ldots b_3 b_2 b_1 b_0$$

Table 3.2 Truth table for subtraction of complex binary numbers

Inputs		Outputs	
Minuend	Subtrahend	Borrows	Difference
0	0	000	0
0	1	1110	1
1	0	000	1
1	1	000	0

Then the result of subtracting 1 from 0 is obtained by changing:

$$a_k \rightarrow a_k + 1 \tag{3.1}$$

$$a_{k+1} \rightarrow a_{k+1} \,(\text{unchanged}) \tag{3.2}$$

$$a_{k+2} \rightarrow a_{k+2} + 1 \tag{3.3}$$

$$a_{k+3} \rightarrow a_{k+3} + 1 \tag{3.4}$$

$$a_{k+4} \rightarrow a_{k+4} + 1 \tag{3.5}$$

$$b_k \rightarrow 0 \tag{3.6}$$

Note that addition of "1" in the above equations is accomplished by the application of addition algorithm described in Sect. 3.1.

As an example, let us subtract $(1 + j3)$ from 2 in CBNS:

$(1 + j3)$ is equivalent to $(1010)_{\text{Base } (-1+j)}$ and 2 is equivalent to $(1100)_{\text{Base } (-1+j)}$. Thus

$$2 - (1 + j3)$$

$$= 1100_{\text{Base } (-1+j)} - 1010_{\text{Base } (-1+j)}$$

Isolating the special case,

$$= 0100_{\text{Base } (-1+j)} - 0010_{\text{Base } (-1+j)}$$

$$= 111110_{\text{Base } (-1+j)} - 000000_{\text{Base } (-1+j)} \text{ (by algorithm)}$$

$$= 111110_{\text{Base } (-1+j)} = 1 - j3$$

3.3 Multiplication Algorithm for Complex Binary Numbers

The binary multiplication of two complex binary numbers (multiplicand and multiplier) consists of two operations [1–3]. First, a bit-wise logical AND operation is carried out between each multiplier bit and the whole multiplicand. This will result in intermediate products, just like in the decimal or binary multiplication process, with each intermediate result shifted one-bit to the left compared to the previous intermediate result. Then, these intermediate products are added according to the CBNS addition algorithm described in Sect. 3.1. The *zero rule* plays an important role in reducing the number of summands resulting from intermediate multiplications.

As an example, let us multiply $(1 + j2)$ with $(2 - j)$

$(1 + j2)$ is equivalent to $(1110101)_{\text{Base } (-1+j)}$ and $(2 - j)$ is equivalent to $(111011)_{\text{Base } (-1+j)}$. Thus

$$(1 + j2) \times (2 - j)$$

$$= (1110101)_{\text{Base } (-1+j)} \times (111011)_{\text{Base } (-1+j)}$$

First, we obtain the intermediate products by taking each multiplier's bit and ANDing it with the multiplicand as follows:

$$
\begin{array}{r}
1110101 \\
\times 111011 \\
\hline
\end{array}
$$

$$
\begin{array}{ll}
1110101 & \\
11101010 & \leftarrow \text{ shifted left by 1-bit} \\
000000000 & \leftarrow\leftarrow \text{ shifted left by 2-bits} \\
1110101000 & \leftarrow\leftarrow\leftarrow \text{ shifted left by 3-bits} \\
11101010000 & \leftarrow\leftarrow\leftarrow\leftarrow \text{ shifted left by 4-bits} \\
111010100000 & \leftarrow\leftarrow\leftarrow\leftarrow\leftarrow \text{ shifted left by 5-bits}
\end{array}
$$

Then, we are going to add all intermediate products according to the addition algorithm described in Sect. 3.1. To reduce the number of addition operations, we'll also identify the operands which satisfy *zero rule* $(111 + 11 = 0)$ by **bold-facing 1s**.

$$
\begin{array}{r}
\mathbf{1110101} \\
\mathbf{1110}1010 \\
000000000 \\
\mathbf{1110101000} \\
\mathbf{1110}1010000 \\
\mathbf{1110}10100000 \\
\hline
1100111
\end{array}
$$

Thus,

$$(1 + j2) \times (2 - j) = (1100111)_{\text{Base } (-1+j)} = (4 + j3)$$

3.4 Division Algorithm for Complex Binary Numbers

Algorithmically, the division operation can be implemented as a multiplication operation between the reciprocal of the divisor (denominator) and the dividend (numerator) [1–3]. For a more detailed mathematical analysis of the division algorithm in CBNS, the reader is referred to [4, 5].

Thus,

$$\frac{(a+jb)}{(c+jd)} = (a+jb)(c+jd)^{-1} = (a+jb)z \tag{3.7}$$

where $z = w^{-1}$ and $w = (c+jd)$.

We start with our initial approximation of z by setting $z_0 = (-1+j)^{-k}$ where k is obtained from the representation of w such that:

$$w = \sum_{i=-\infty}^{k} a_i(-1+j)^{-i} \tag{3.8}$$

in which $a_k \equiv 1$ and $a_i \equiv 0$ for $i > k$. The successive approximations are then obtained by $z_{i+1} = z_i(2 - wz_i)$. If the values of z do not converge, we correct our initial approximation of z by setting $z_0 = j(-1+j)^{-k}$ which will definitely converge [4, 5]. Having calculated the value of z, we can multiply it with $(a+jb)$ to obtain the result of the division operation.

In the following examples, for the sake of clarity, we have used decimal number system for the successive values of z to explain the converging process of the division algorithm.

Let us assume that we want to obtain the result of dividing $(1+j2)$ by $(1+j3)$. Thus $(a+jb) = (1+j2)$ and $w = (c+jd) = (1+j3)$. Our calculations for approximation of $z = w^{-1}$ begin by first determining the value of k as follows:

$$w = (1+j3) = 1010_{\text{Base }(-1+j)}$$

$$= 1 \times (-1+j)^3 + 0 \times (-1+j)^2 + 1 \times (-1+j)^1 + 0 \times (-1+j)^0 \Rightarrow k = 3$$

Therefore,

$$z_0 = (-1+j)^{-k} = (-1+j)^{-3} = 0.25 - j0.25$$

$$z_1 = z_0(2 - wz_0) = 0.125 - j0.375$$

$$z_2 = z_1(2 - wz_1) = 0.09375 - j0.28125$$

$$z_3 = z_2(2 - wz_2) = 0.099609375 - j0.2988281250$$

$$z_4 = z_3(2 - wz_3) = 0.09999847412 - j0.2999954224$$

$$z_5 = z_4(2 - wz_4) = 0.1 - j0.3$$

$$z_6 = z_5(2 - wz_5) = 0.1 - j0.3$$

Now

$$(0.1 - j0.3) = (0.00111100101111001011100\ldots)_{\text{Base } (-1+j)}$$

So

$$\frac{1+j2}{1+j3} = (1+j2) \times (1+j3)^{-1}$$

$$= (1110101)_{\text{Base } (-1+j)}$$

$$\times (0.00111100101111001011100\ldots)_{\text{Base } (-1+j)}$$

$$= (1.11111001011110010111110010111\ldots)_{\text{Base } (-1+j)}$$

$$= 0.7 - j0.1$$

As another example, let $w = (-28 - j15)$
Then

$$(-28 - j15) = (11011010011)_{\text{Base } (-1+j)}$$

$$= 1 \times (-1+j)^{10} + 1 \times (-1+j)^9 + 0$$

$$\times (-1+j)^8 + 1 \times (-1+j)^7 + 1$$

$$\times (-1+j)^6 + 0 \times (-1+j)^5 + 1$$

$$\times (-1+j)^4 + 0 \times (-1+j)^3 + 0$$

$$\times (-1+j)^2 + 1 \times (-1+j)^1 + 1$$

$$\times (-1+j)^0 \Rightarrow k = 10$$

We begin by choosing

$$z_0 = (-1+j)^{-k} = (-1+j)^{-10} = j0.03125$$

$$z_1 = z_0(2 - wz_0) = -0.02734375 + j0.0478515625$$

$$z_2 = z_1(2 - wz_1) = -0.05861282347 - j0.0007009506465$$

$$z_3 = z_2(2 - wz_2) = -0.02227897905 + j0.05242341786$$

$$z_4 = z_3(2 - wz_3) = -0.07257188256 + j0.005664255473$$

$$z_5 = z_4(2 - wz_4) = 0.01375684714 + j0.06682774792$$

$$z_6 = z_5(2 - wz_5) = -0.1198139965 + j0.1209880304$$

$$z_7 = z_6(2 - wz_6) = 0.1873379177 - j0.5740439146$$

$$z_8 = z_7(2 - wz_7) = -4.643184188 - j11.58680236$$

$$z_9 = z_8(2 - wz_8) = -4778.731320 + j1299.184773$$

As evident from the above calculations, the value of z is not converging with each successive iteration. Therefore, we correct our initial approximation to:

$$z_0 = j(-1 + j)^{-k} = j(-1 + j)^{-10} = -0.03125$$

$$z_1 = z_0(2 - wz_0) = -0.03515625 + j0.0146484375$$

$$z_2 = z_1(2 - wz_1) = -0.02626419066 + j0.01577854156$$

$$z_3 = z_2(2 - wz_2) = -0.02775239316 + j0.01496276824$$

$$z_4 = z_3(2 - wz_3) = -0.02775050252 + j0.01486605276$$

$$z_5 = z_4(2 - wz_4) = -0.02775024777 + j0.01486620416$$

$$z_6 = z_5(2 - wz_5) = -0.02775024777 + j0.01486620416$$

$$z_7 = z_6(2 - wz_6) = -0.02775024777 + j0.01486620416$$

$$z_8 = z_7(2 - wz_7) = -0.02775024777 + j0.01486620416$$

$$z_9 = z_8(2 - wz_8) = -0.02775024777 + j0.01486620416$$

As is clear from the above calculations, the value of z is converging to $(-0.0277 + j0.0148)$. This value of z can be multiplied with any given complex number (numerator) to obtain the result of dividing the complex number by $(-28 - j15)$, as in previous example.

3.5 Effect of Shift-Left Operations on Complex Binary Numbers

To investigate the effects of shift-left (1, 2, 3, and 4-bits) operations on a complex number represented in CBNS format, a computer program was developed in C++ language which allowed (i) variations in magnitude and sign of both real and imaginary components of a complex number to be generated automatically in a linear fashion, and (ii) decomposition of the complex number after the shift-left operation, represented in CBNS format, into its real and imaginary components [6–8]. The length of the original binary bit array was restricted to 800 bits and 0s were padded on the left side of the binary data when the given complex number required less than maximum allowable bits for representation in CBNS format before the shift operation. As an example, consider the following complex number:

Before padding

$$(90 + j90)_{10} = (110100010001000)_{\text{Base } (-1+j)}$$

After padding

$$(90 + j90)_{10} = (0\ldots0110100010001000)_{\text{Base } (-1+j)}$$

such that the total size of the binary array is 800 bits. Shifting this binary array by 1-bit to the left will yield $(0\ldots01101000100010000)_{\text{Base } (-1+j)}$ by removing one 0 from the left side and appending it to the right side of the number, ensuring that total array size remains 800 bits. Similarly, shifting of the original binary array by 2, 3, or 4-bits to the left will yield $(0\ldots011010001000100000)_{\text{Base } (-1+j)}$ (notice two 0s appended on the right-side of the array), $(0\ldots01101000\ 10001000000)_{\text{Base } (-1+j)}$ (notice three 0s appended on the right-side of the array), $(0\ldots01101000100010000000)_{\text{Base } (-1+j)}$ (notice four 0s appended on the right-side of the array), ensuring all the time that total arraysize remains 800 bits by removing two 0s, three 0s, and four 0s, respectively, from the left side of the original array.

Table 3.3 presents characteristic equations describing complex numbers in CBNS format after shift-left operations. We have assumed that initially our complex number is given by $\pm Real_{old} \pm jImag_{old}$ and, after shift-left operation by 1, 2, 3, or 4-bits is represented by $\pm Real_{new} \pm jImag_{new}$.

Table 3.3 Characteristic equations describing complex numbers in CBNS format after shift-left operations [7, 8]

Type of complex number	Shift-left by 1-bit		Shift-left by 2-bits	
	$Real_{new}$	$Imag_{new}$	$Real_{new}$	$Imag_{new}$
Positive Real	$-Real_{old}$	$+Real_{old}$	0	$-2Real_{old}$
Negative Real	$-Real_{old}$	$+Real_{old}$	0	$-2Real_{old}$
Positive Imag	$-Imag_{old}$	$-Imag_{old}$	$+2Imag_{old}$	0
Negative Imag	$-Imag_{old}$	$-Imag_{old}$	$+2Imag_{old}$	0
+Real +Imag	$-2Real_{old}$	0	$+2Real_{old}$	$-2Imag_{old}$
+Real −Imag	0	$-2Imag_{old}$	$-2Real_{old}$	$+2Imag_{old}$
−Real +Imag	0	$-2Imag_{old}$	$-2Real_{old}$	$+2Imag_{old}$
−Real −Imag	$-2Real_{old}$	0	$+2Real_{old}$	$-2Imag_{old}$
Type of complex number	Shift-left by 3-bits		Shift-left by 4-bits	
	$Real_{new}$	$Imag_{new}$	$Real_{new}$	$Imag_{new}$
Positive Real	$+2Real_{old}$	$+2Real_{old}$	$-4Real_{old}$	0
Negative Real	$+2Real_{old}$	$+2Real_{old}$	$-4Real_{old}$	0
Positive Imag	$-2Imag_{old}$	$+2Imag_{old}$	0	$-4Imag_{old}$
Negative Imag	$-2Imag_{old}$	$+2Imag_{old}$	0	$-4Imag_{old}$
+Real +Imag	0	$+4Imag_{old}$	$-4Real_{old}$	$-4Imag_{old}$
+Real −Imag	$+4Real_{old}$	0	$-4Real_{old}$	$-4Imag_{old}$
−Real +Imag	$+4Real_{old}$	0	$-4Rea_{old}$	$-4Imag_{old}$
−Real −Imag	$+4Real_{old}$	0	$-4Real_{old}$	$-4Imag_{old}$

3.6 Effect of Shift-Right Operations on Complex Binary Numbers

To investigate the effects of shift-right (1, 2, 3, and 4-bits) operations on a complex number represented in CBNS format, a computer program, similar to the one used for shift-left operations, was developed in C++ language which allowed (i) variations in magnitude and sign of both real and imaginary components of a complex number to be generated automatically in a linear fashion, and (ii) decomposition of the complex number after the shift-right operation, represented in CBNS format, into its real and imaginary components [8–10]. The length of the original binary bit array was restricted to 800 bits, as before for shift-left operations, and 0s were padded on the left-side of the binary data when the given complex number required less than maximum allowable bits for representation in CBNS format. Shifting the padded binary string by 1, 2, 3, and 4-bits to the right caused one, two, three, and four 0s to be inserted on the left-side of the string such that the total length of the string remained 800 bits.

Table 3.4 presents characteristic equations describing complex numbers in CBNS format after shift-right operations. Here, again we have assumed that initially our complex number is given by $\pm Real_{old} \pm jImag_{old}$ and, after shift-right operation by 1, 2, 3, or 4-bits is represented by $\pm Real_{new} \pm jImag_{new}$.

Table 3.4 Characteristic equations describing complex numbers in CBNS format after shift-right operations [8, 10]

Type of complex number	Shift-right by 1-bit		Shift-right by 2-bits	
	$Real_{new}$	$Imag_{new}$	$Real_{new}$	$Imag_{new}$
Positive Real	$-1/2\ Real_{old} +1/4$	$-1/2\ Real_{old}+1/4$	0	$+1/2\ Real_{old}-1/4$
Negative Real	$-1/2\ Real_{old}+1/4$	$-1/2\ Real_{old}+1/4$	0	$+1/2\ Real_{old}-1/4$
Positive Imag	$1/2\ Imag_{old}+1/4$	$-1/2\ Imag_{old}+1/4$	$-1/2\ Imag_{old}+1/4$	0
Negative Imag	$1/2\ Imag_{old}+1/4$	$-1/2\ Imag_{old}+1/4$	$-1/2\ Imag_{old}+1/4$	0
+Real +Imag	0	$-\ Imag_{old}$	$-1/2\ Real_{old}+1/4$	$1/2\ Imag_{old}+1/4$
+Real −Imag	$-Real_{old}$	0	$1/2\ Real_{old}+1/4$	$1/2\ Imag_{old}+1/4$
−Real +Imag	$Real_{old}$	0	$1/2\ Real_{old}+1/4$	$-1/2\ Imag_{old}+1/4$
−Real −Imag	0	$-Imag_{old}$	$-1/2\ Real_{old}+1/4$	$1/2\ Imag_{old}+1/4$

Type of complex number	Shift-right by 3-bits		Shift-right by 4-bits	
	$Real_{new}$	$Imag_{new}$	$Real_{new}$	$Imag_{new}$
Positive Real	$1/4\ Real_{old}$	$-1/4\ Real_{old}$	$-1/4\ Real_{old}$	0
Negative Real	$1/4\ Real_{old}$	$-1/4\ Real_{old}$	$-1/4\ Real_{old}$	0
Positive Imag	$1/4\ Imag_{old}$	$1/4\ Imag_{old}$	0	$-1/4\ Imag_{old}$
Negative Imag	$1/4\ Imag_{old}$	$1/4\ Imag_{old}$	0	$-1/4\ Imag_{old}$
+Real +Imag	$1/2\ Real_{old}+1/4$	0	$-1/4\ Real_{old}$	$-1/4\ Imag_{old}$
+Real −Imag	0	$1/2\ Imag_{old}-1/4$	$-1/4\ Real_{old}$	$-1/4\ Imag_{old}$
−Real +Imag	0	$1/2\ Imag_{old}-1/4$	$-1/4\ Real_{old}$	$-1/4\ Imag_{old}$
−Real −Imag	$1/2\ Real_{old}+1/4$	0	$-1/4\ Real_{old}$	$-1/4\ Imag_{old}$

References

1. T. Jamil, N. Holmes, D. Blest, Towards implementation of a binary number system for complex numbers, in *Proceedings of the IEEE Southeastcon*, pp. 268–274 (2000)
2. T. Jamil, The complex binary number system: Basic arithmetic made simple. IEEE Potentials **20**(5), 39–41 (2002)
3. T. Jamil, An introduction to complex binary number system, in *Proceedings of the 4th International Conference on Information and Computing*, pp. 229–232 (2011)
4. D. Blest, T. Jamil, Efficient division in the binary representation of complex numbers, in *Proceedings of the IEEE Southeastcon*, pp. 188–195 (2001)
5. D. Blest, T. Jamil, Division in a binary representation for complex numbers. Int. J. Math. Educ. Sci. Tech. **34**(4), 561–574 (2003)
6. T. Jamil, U. Ali, Effects of shift-left operations on complex binary numbers, in *Proceedings of the 18th Annual Canadian Conference on Electrical and Computer Engineering*, pp. 1951–1954 (2005)
7. T. Jamil, U. Ali, An investigation into the effects of multiple-bit shift-left operations on $(-1+j)$-base representation of complex numbers, in *Proceedings of the International Conference on Computer and Communication Engineering* (1), pp. 549–554 (2006)
8. T. Jamil, Impact of shift operations on $(-1+j)$-base complex binary numbers. J. Comput. **3**(2), 63–71 (2008)
9. U. Ali, T. Jamil, S. Saeed, Effects of shift-right operations on binary representation of complex numbers, in *Proceedings of the International Conference on Communication, Computer and Power*, pp. 238–243 (2005)
10. T. Jamil, U. Ali, Effects of multiple-bit shift-right operations on complex binary numbers, in *Proceedings of the IEEE SoutheastCon*, pp. 759–764 (2007)

Chapter 4
Arithmetic Circuits Designs

Abstract The algorithms for arithmetic operations in CBNS, described in the previous chapter, have been implemented in computer hardware using Field Programmable Gate Arrays (FPGAs). This chapter includes design information for a nibble-size (four bits) adder, subtractor, multiplier, and divider circuits utilizing CBNS for representation of complex numbers. The implementation and performance statistics related to these circuits are also presented.

4.1 Adder Circuit for Complex Binary Numbers

There have been three known designs for CBNS-based adder circuits published in the scientific literature [1–5]. These circuits have been based on the concepts of minimum-delay, ripple-carry, and state-machine.

4.1.1 Minimum-Delay Adder

The minimum-delay nibble-size CBNS adder has been designed by first writing a truth table with four-bit augend ($a_3a_2a_1a_0$) and addend ($b_3b_2b_1b_0$) operands as inputs (total of $2^8 = 256$ minterms), and twelve outputs ($c_{11}c_{10}c_9c_8c_7c_6c_5c_4s_3s_2s_1s_0$) which are obtained by adding each pair of nibble-size inputs according to the addition algorithm described in Chap. 3. Each output is then expressed in sum-of-minterms form. The resulting design expressions have been implemented using an 8×256 decoder (to generate each minterm) and multiple-input OR gates (to combine relevant minterms for each output).

Tables 4.1, 4.2, 4.3, 4.4, 4.5, 4.6, 4.7, 4.8 present complete truth table for a nibble-size minimum-delay complex binary adder. For the sake of simplicity, the

T. Jamil, *Complex Binary Number System*,
SpringerBriefs in Electrical and Computer Engineering,
DOI: 10.1007/978-81-322-0854-9_4, © The Author(s) 2013

Table 4.1 Truth table for a nibble-size minimum-delay complex binary adder [1, 3, 5] (Minterm: $a_3a_2a_1a_0$ ADD $b_3b_2b_1b_0 = c_{11}c_{10}c_9c_8c_7c_6c_5c_4s_3s_2s_1s_0$)

Minterm	Augend				Addend				Sum
	a_3	a_2	a_1	a_0	b_3	b_2	b_1	b_0	$c_{11}c_{10}...s_1s_0$
0	0	0	0	0	0	0	0	0	000000000000
1	0	0	0	0	0	0	0	1	000000000001
2	0	0	0	0	0	0	1	0	000000000010
3	0	0	0	0	0	0	1	1	000000000011
4	0	0	0	0	0	1	0	0	000000000100
5	0	0	0	0	0	1	0	1	000000000101
6	0	0	0	0	0	1	1	0	000000000110
7	0	0	0	0	0	1	1	1	000000000111
8	0	0	0	0	1	0	0	0	000000001000
9	0	0	0	0	1	0	0	1	000000001001
10	0	0	0	0	1	0	1	0	000000001010
11	0	0	0	0	1	0	1	1	000000001011
12	0	0	0	0	1	1	0	0	000000001100
13	0	0	0	0	1	1	0	1	000000001101
14	0	0	0	0	1	1	1	0	000000001110
15	0	0	0	0	1	1	1	1	000000001111
16	0	0	0	1	0	0	0	0	000000000001
17	0	0	0	1	0	0	0	1	000000001100
18	0	0	0	1	0	0	1	0	000000000011
19	0	0	0	1	0	0	1	1	000000001110
20	0	0	0	1	0	1	0	0	000000000101
21	0	0	0	1	0	1	0	1	000000111000
22	0	0	0	1	0	1	1	0	000000000111
23	0	0	0	1	0	1	1	1	000000111010
24	0	0	0	1	1	0	0	0	000000001001
25	0	0	0	1	1	0	0	1	000001100100
26	0	0	0	1	1	0	1	0	000000001011
27	0	0	0	1	1	0	1	1	000001100110
28	0	0	0	1	1	1	0	0	000000001101
29	0	0	0	1	1	1	0	1	000111010000
30	0	0	0	1	1	1	1	0	000000001111
31	0	0	0	1	1	1	1	1	000111010010

twelve outputs have been collectively labeled as "Sum" in these tables. The sum-of-minterms expressions for outputs of the adder are listed in Tables 4.9, 4.10. Block diagram of the adder is given in Fig. 4.1.

Table 4.2 Truth table for a nibble-size minimum-delay complex binary adder [1, 3, 5] (Minterm: $a_3a_2a_1a_0$ ADD $b_3b_2b_1b_0 = c_{11}c_{10}c_9c_8c_7c_6c_5c_4s_3s_2s_1s_0$)

Minterm	Augend				Addend				Sum
	a_3	a_2	a_1	a_0	b_3	b_2	b_1	b_0	$c_{11}c_{10}...s_1s_0$
32	0	0	1	0	0	0	0	0	000000000010
33	0	0	1	0	0	0	0	1	000000000011
34	0	0	1	0	0	0	1	0	000000011000
35	0	0	1	0	0	0	1	1	000000011001
36	0	0	1	0	0	1	0	0	000000000110
37	0	0	1	0	0	1	0	1	000000000111
38	0	0	1	0	0	1	1	0	000000011100
39	0	0	1	0	0	1	1	1	000000011101
40	0	0	1	0	1	0	0	0	000000001010
41	0	0	1	0	1	0	0	1	000000001011
42	0	0	1	0	1	0	1	0	000001110000
43	0	0	1	0	1	0	1	1	000001110001
44	0	0	1	0	1	1	0	0	000000001110
45	0	0	1	0	1	1	0	1	000000001111
46	0	0	1	0	1	1	1	0	000001110100
47	0	0	1	0	1	1	1	1	000001110101
48	0	0	1	1	0	0	0	0	000000000011
49	0	0	1	1	0	0	0	1	000000001110
50	0	0	1	1	0	0	1	0	000000011001
51	0	0	1	1	0	0	1	1	000001110100
52	0	0	1	1	0	1	0	0	000000000111
53	0	0	1	1	0	1	0	1	000000111010
54	0	0	1	1	0	1	1	0	000000011101
55	0	0	1	1	0	1	1	1	000000000000
56	0	0	1	1	1	0	0	0	000000001011
57	0	0	1	1	1	0	0	1	000001100110
58	0	0	1	1	1	0	1	0	000001110001
59	0	0	1	1	1	0	1	1	000001111100
60	0	0	1	1	1	1	0	0	000000001111
61	0	0	1	1	1	1	0	1	000111010010
62	0	0	1	1	1	1	1	0	000001110101
63	0	0	1	1	1	1	1	1	000000001000

Table 4.3 Truth table for a nibble-size minimum-delay complex binary adder [1, 3, 5] (Minterm: $a_3a_2a_1a_0$ ADD $b_3b_2b_1b_0 = c_{11}c_{10}c_9c_8c_7c_6c_5c_4s_3s_2s_1s_0$)

Minterm	Augend				Addend				Sum
	a_3	a_2	a_1	a_0	b_3	b_2	b_1	b_0	$c_{11}c_{10}...s_1s_0$
64	0	1	0	0	0	0	0	0	000000000100
65	0	1	0	0	0	0	0	1	000000000101
66	0	1	0	0	0	0	1	0	000000000110
67	0	1	0	0	0	0	1	1	000000000111
68	0	1	0	0	0	1	0	0	000000110000
69	0	1	0	0	0	1	0	1	000000110001
70	0	1	0	0	0	1	1	0	000000110010
71	0	1	0	0	0	1	1	1	000000110011
72	0	1	0	0	1	0	0	0	000000001100
73	0	1	0	0	1	0	0	1	000000001101
74	0	1	0	0	1	0	1	0	000000001110
75	0	1	0	0	1	0	1	1	000000001111
76	0	1	0	0	1	1	0	0	000000111000
77	0	1	0	0	1	1	0	1	000000111001
78	0	1	0	0	1	1	1	0	000000111010
79	0	1	0	0	1	1	1	1	000000111011
80	0	1	0	1	0	0	0	0	000000000101
81	0	1	0	1	0	0	0	1	000000111000
82	0	1	0	1	0	0	1	0	000000000111
83	0	1	0	1	0	0	1	1	000000111010
84	0	1	0	1	0	1	0	0	000000110001
85	0	1	0	1	0	1	0	1	000000111100
86	0	1	0	1	0	1	1	0	000000110011
87	0	1	0	1	0	1	1	1	000000111110
88	0	1	0	1	1	0	0	0	000000001101
89	0	1	0	1	1	0	0	1	000111010000
90	0	1	0	1	1	0	1	0	000000001111
91	0	1	0	1	1	0	1	1	000111010010
92	0	1	0	1	1	1	0	0	000000111001
93	0	1	0	1	1	1	0	1	000111010100
94	0	1	0	1	1	1	1	0	000000111011
95	0	1	0	1	1	1	1	1	000111010110

Table 4.4 Truth table for a nibble-size minimum-delay complex binary adder [1, 3, 5] (Minterm: $a_3a_2a_1a_0$ ADD $b_3b_2b_1b_0 = c_{11}c_{10}c_9c_8c_7c_6c_5c_4s_3s_2s_1s_0$)

Minterm	Augend				Addend				Sum
	a_3	a_2	a_1	a_0	b_3	b_2	b_1	b_0	$c_{11}c_{10}...s_1s_0$
96	0	1	1	0	0	0	0	0	000000000110
97	0	1	1	0	0	0	0	1	000000000111
98	0	1	1	0	0	0	1	0	000000011100
99	0	1	1	0	0	0	1	1	000000011101
100	0	1	1	0	0	1	0	0	000000110010
101	0	1	1	0	0	1	0	1	000000110011
102	0	1	1	0	0	1	1	0	000011101000
103	0	1	1	0	0	1	1	1	000011101001
104	0	1	1	0	1	0	0	0	000000001110
105	0	1	1	0	1	0	0	1	000000001111
106	0	1	1	0	1	0	1	0	000001110100
107	0	1	1	0	1	0	1	1	000001110101
108	0	1	1	0	1	1	0	0	000000111010
109	0	1	1	0	1	1	0	1	000000111011
110	0	1	1	0	1	1	1	0	000000000000
111	0	1	1	0	1	1	1	1	000000000001
112	0	1	1	1	0	0	0	0	000000000111
113	0	1	1	1	0	0	0	1	000000111010
114	0	1	1	1	0	0	1	0	000000011101
115	0	1	1	1	0	0	1	1	000000000000
116	0	1	1	1	0	1	0	0	000000110011
117	0	1	1	1	0	1	0	1	000000111110
118	0	1	1	1	0	1	1	0	000011101001
119	0	1	1	1	0	1	1	1	000000000100
120	0	1	1	1	1	0	0	0	000000001111
121	0	1	1	1	1	0	0	1	000111010010
122	0	1	1	1	1	0	1	0	000001110101
123	0	1	1	1	1	0	1	1	000000001000
124	0	1	1	1	1	1	0	0	000000111011
125	0	1	1	1	1	1	0	1	000111010110
126	0	1	1	1	1	1	1	0	000000000001
127	0	1	1	1	1	1	1	1	000000001100

Table 4.5 Truth table for a nibble-size minimum-delay complex binary adder [1, 3, 5] (Minterm: $a_3a_2a_1a_0$ ADD $b_3b_2b_1b_0 = c_{11}c_{10}c_9c_8c_7c_6c_5c_4s_3s_2s_1s_0$)

Minterm	Augend				Addend				Sum
	a_3	a_2	a_1	a_0	b_3	b_2	b_1	b_0	$c_{11}c_{10}\ldots s_1s_0$
128	1	0	0	0	0	0	0	0	000000001000
129	1	0	0	0	0	0	0	1	000000001001
130	1	0	0	0	0	0	1	0	000000001010
131	1	0	0	0	0	0	1	1	000000001011
132	1	0	0	0	0	1	0	0	000000001100
133	1	0	0	0	0	1	0	1	000000001101
134	1	0	0	0	0	1	1	0	000000001110
135	1	0	0	0	0	1	1	1	000000001111
136	1	0	0	0	1	0	0	0	000001100000
137	1	0	0	0	1	0	0	1	000001100001
138	1	0	0	0	1	0	1	0	000001100010
139	1	0	0	0	1	0	1	1	000001100011
140	1	0	0	0	1	1	0	0	000001100100
141	1	0	0	0	1	1	0	1	000001100101
142	1	0	0	0	1	1	1	0	000001100110
143	1	0	0	0	1	1	1	1	000001100111
144	1	0	0	1	0	0	0	0	000000001001
145	1	0	0	1	0	0	0	1	000001100100
146	1	0	0	1	0	0	1	0	000000001011
147	1	0	0	1	0	0	1	1	000001100110
148	1	0	0	1	0	1	0	0	000000001101
149	1	0	0	1	0	1	0	1	000111010000
150	1	0	0	1	0	1	1	0	000000001111
151	1	0	0	1	0	1	1	1	000111010010
152	1	0	0	1	1	0	0	0	000001100001
153	1	0	0	1	1	0	0	1	000001101100
154	1	0	0	1	1	0	1	0	000001100011
155	1	0	0	1	1	0	1	1	000001101110
156	1	0	0	1	1	1	0	0	000001100101
157	1	0	0	1	1	1	0	1	000111011000
158	1	0	0	1	1	1	1	0	000001100111
159	1	0	0	1	1	1	1	1	000111011010

Table 4.6 Truth table for a nibble-size minimum-delay complex binary adder [1, 3, 5] (Minterm: $a_3a_2a_1a_0$ ADD $b_3b_2b_1b_0 = c_{11}c_{10}c_9c_8c_7c_6c_5c_4s_3s_2s_1s_0$)

Minterm	Augend				Addend				Sum
	a_3	a_2	a_1	a_0	b_3	b_2	b_1	b_0	$c_{11}c_{10}...s_1s_0$
160	1	0	1	0	0	0	0	0	000000001010
161	1	0	1	0	0	0	0	1	000000001011
162	1	0	1	0	0	0	1	0	000001110000
163	1	0	1	0	0	0	1	1	000001110001
164	1	0	1	0	0	1	0	0	000000001110
165	1	0	1	0	0	1	0	1	000000001111
166	1	0	1	0	0	1	1	0	000001110100
167	1	0	1	0	0	1	1	1	000001110101
168	1	0	1	0	1	0	0	0	000001100010
169	1	0	1	0	1	0	0	1	000001100011
170	1	0	1	0	1	0	1	0	000001111000
171	1	0	1	0	1	0	1	1	000001111001
172	1	0	1	0	1	1	0	0	000001100110
173	1	0	1	0	1	1	0	1	000001100111
174	1	0	1	0	1	1	1	0	000001111100
175	1	0	1	0	1	1	1	1	000001111101
176	1	0	1	1	0	0	0	0	000000001011
177	1	0	1	1	0	0	0	1	000001100110
178	1	0	1	1	0	0	1	0	000001110001
179	1	0	1	1	0	0	1	1	000001111100
180	1	0	1	1	0	1	0	0	000000001111
181	1	0	1	1	0	1	0	1	000111010010
182	1	0	1	1	0	1	1	0	000001110101
183	1	0	1	1	0	1	1	1	000000001000
184	1	0	1	1	1	0	0	0	000001100011
185	1	0	1	1	1	0	0	1	000001101110
186	1	0	1	1	1	0	1	0	000001111001
187	1	0	1	1	1	0	1	1	111010010100
188	1	0	1	1	1	1	0	0	000001100111
189	1	0	1	1	1	1	0	1	000111011010
190	1	0	1	1	1	1	1	0	000001111101
191	1	0	1	1	1	1	1	1	000001100000

Table 4.7 Truth table for a nibble-size minimum-delay complex binary adder [1, 3, 5] (Minterm: $a_3a_2a_1a_0$ ADD $b_3b_2b_1b_0 = c_{11}c_{10}c_9c_8c_7c_6c_5c_4s_3s_2s_1s_0$)

Minterm	Augend				Addend				Sum
	a_3	a_2	a_1	a_0	b_3	b_2	b_1	b_0	$c_{11}c_{10}\ldots s_1s_0$
192	1	1	0	0	0	0	0	0	000000001100
193	1	1	0	0	0	0	0	1	000000001101
194	1	1	0	0	0	0	1	0	000000001110
195	1	1	0	0	0	0	1	1	000000001111
196	1	1	0	0	0	1	0	0	000000111000
197	1	1	0	0	0	1	0	1	000000111001
198	1	1	0	0	0	1	1	0	000000111010
199	1	1	0	0	0	1	1	1	000000111011
200	1	1	0	0	1	0	0	0	000001100100
201	1	1	0	0	1	0	0	1	000001100101
202	1	1	0	0	1	0	1	0	000001100110
203	1	1	0	0	1	0	1	1	000001100111
204	1	1	0	0	1	1	0	0	000111010000
205	1	1	0	0	1	1	0	1	000111010001
206	1	1	0	0	1	1	1	0	000111010010
207	1	1	0	0	1	1	1	1	000111010011
208	1	1	0	1	0	0	0	0	000000001101
209	1	1	0	1	0	0	0	1	000111010000
210	1	1	0	1	0	0	1	0	000000001111
211	1	1	0	1	0	0	1	1	000111010010
212	1	1	0	1	0	1	0	0	000000111001
213	1	1	0	1	0	1	0	1	000111010100
214	1	1	0	1	0	1	1	0	000000111011
215	1	1	0	1	0	1	1	1	000111010110
216	1	1	0	1	1	0	0	0	000001100101
217	1	1	0	1	1	0	0	1	000111011000
218	1	1	0	1	1	0	1	0	000001100111
219	1	1	0	1	1	0	1	1	000111011010
220	1	1	0	1	1	1	0	0	000111010001
221	1	1	0	1	1	1	0	1	000111011100
222	1	1	0	1	1	1	1	0	000111010011
223	1	1	0	1	1	1	1	1	000111011110

Table 4.8 Truth table for a nibble-size minimum-delay complex binary adder [1, 3, 5] (Minterm: $a_3a_2a_1a_0$ ADD $b_3b_2b_1b_0 = c_{11}c_{10}c_9c_8c_7c_6c_5c_4s_3s_2s_1s_0$)

Minterm	Augend				Addend				Sum
	a_3	a_2	a_1	a_0	b_3	b_2	b_1	b_0	$c_{11}c_{10}...s_1s_0$
224	1	1	1	0	0	0	0	0	000000001110
225	1	1	1	0	0	0	0	1	000000001111
226	1	1	1	0	0	0	1	0	000001110100
227	1	1	1	0	0	0	1	1	000001110101
228	1	1	1	0	0	1	0	0	000000111010
229	1	1	1	0	0	1	0	1	000000111011
230	1	1	1	0	0	1	1	0	000000000000
231	1	1	1	0	0	1	1	1	000000000001
232	1	1	1	0	1	0	0	0	000001100110
233	1	1	1	0	1	0	0	1	000001100111
234	1	1	1	0	1	0	1	0	000001111100
235	1	1	1	0	1	0	1	1	000001111101
236	1	1	1	0	1	1	0	0	000111010010
237	1	1	1	0	1	1	0	1	000111010011
238	1	1	1	0	1	1	1	0	000000001000
239	1	1	1	0	1	1	1	1	000000001001
240	1	1	1	1	0	0	0	0	000000001111
241	1	1	1	1	0	0	0	1	000111010010
242	1	1	1	1	0	0	1	0	000001110101
243	1	1	1	1	0	0	1	1	000000001000
244	1	1	1	1	0	1	0	0	000000111011
245	1	1	1	1	0	1	0	1	000111010110
246	1	1	1	1	0	1	1	0	000000000001
247	1	1	1	1	0	1	1	1	000000001100
248	1	1	1	1	1	0	0	0	000001100111
249	1	1	1	1	1	0	0	1	000111011010
250	1	1	1	1	1	0	1	0	000001111101
251	1	1	1	1	1	0	1	1	000001100000
252	1	1	1	1	1	1	0	0	000111010011
253	1	1	1	1	1	1	0	1	000111011110
254	1	1	1	1	1	1	1	0	000000001001
255	1	1	1	1	1	1	1	1	000001100100

Table 4.9 Minterms corresponding to outputs of a nibble-size minimum-delay adder [1, 3, 5]

Adder outputs	Corresponding minterms
c_{11}	187
c_{10}	187
c_9	187
c_8	29,31,61,89,91,93,95,121,125,149,151,157,159,181,189, 204,205,206,207,209,211,213,215,217,219,220,221,222, 223,236,237,241,245,249,252,253
c_7	29,31,61,89,91,93,95,102,103,118,121,125,149,151,157, 159,181,187,189,204,205,206,207,209,211,213,215,217, 219,220,221,222,223,236,237,241,245,249,252,253
c_6	25,27,29,31,42,43,46,47,51,57,58,59,61,62,89,91,93,95, 102,103,106,107,118,121,122,125,136,137,138,139,140, 141,142,143,145,147,149,151,152,153,154,155,156,157, 158,159,162,163,166,167,168,169,170,171,172,173,174, 175,177,178,179,181,182,184,185,186,188,189,190,191, 200,201,202,203,204,205,206,207,209,211,213,215,216, 217,218,219,220,221, 222,223,226,227,232,233,234,235, 236,237,241,242, 245,248,249,250,251,252, 253,255
c_5	21,23,25,27,42,43,46,47,51,53,57,58,59,62,68,69,70,71, 76,77,78,79,81,83,84,85,86,87,92,94,100,101,102,103,106, 107,108,109,113,116,117,118,122,124,136,137,138,139, 140,141,142,143,145,147,152,153,154,155,156,158,162, 163,166,167,168,169,170,171,172,173,174,175,177,178, 179,182,184,185,186,188,190,191,196,197,198,199,200, 201,202,203,212,214,216,218,226,227,228,229,232,233, 234,235,242,244,248,250,251,255
c_4	21,23,29,31,34,35,38,39,42,43,46,47,50,51,53,54,58,59,61, 62,68,69,70,71,76,77,78,79,81,83,84,85,86,87,89,91,92,93, 94,95,98,99,100,101,106,107,108,109,113,114,116,117,121, 122,124,125,149,151,157,159,162,163,166,167,170,171, 174,175,178,179,181,182,186,187,189,190,196,197,198, 199,204,205,206,207,209,211,212,213,214,215,217,219, 220,221,222,223,226,227,228,229,234,235,236,237,241, 242,244,245,249,250,252,253

Table 4.10 Minterms corresponding to Outputs of a Nibble-size Minimum-Delay Adder [1, 3, 5]

Adder outputs	Corresponding minterms
s_3	8,9,10,11,12,13,14,15,17,19,21,23,24,26,28,30,34,35,38, 39,40,41,44,45,49,50,53,54,56,59,60,63,72,73,74,75,76, 77, 78,79,81,83,85,87,88,90,92,94,98,99,102,103,104,105,108, 109,113,114,117,118,120,123,124,127,128,129,130,131, 132,133,134,135,144,146,148,150,153,155,157, 159,160, 161,164,165,170, 171,174,175,176,179,180,183,185,186, 189,190,192, 193,194,195,196,197,198,199,208,210,212, 214,217,219,221,223,224,225,228,229,234,235,238,239, 240,243, 244,247,249, 250,253,254
s_2	4,5,6,7,12,13,14,15,17,19,20,22,25,27,28,30,36,37,38,39, 44,45,46,47,49,51,52,54,57,59,60,62,64,65,66, 67,72,73,74, 75,80,82,85,87,88,90,93,95,96,97,98,99,104,105,106,107, 112,114,117,119,120,122,125,127,132,133,134,135,140, 141,142,143,145,147,148,150,153, 155,156,158,164,165, 166,167,172,173,174,175,177,179,180,182,185,187,188, 190,192,193,194,195,200,201,202,203,208,210,213,215, 216,218,221,223,224,225,226,227,232,233,234,235,240, 242,245, 247,248,250, 253,255
s_1	2,3,6,7,10,11,14,15,18,19,22,23,26,27,30,31,32,33,36,37,40, 41,44,45,48,49,52,53,56,57, 60,61,66,67,70,71,74,75,78, 79,82,83,86,87,90,91,94,95,96,97,100,101,104,105,108,109, 112,113,116,117,120,121,124,125,130,131,134,135,138, 139,142,143,146,147,150,151,154,155,158,159,160,161, 164,165,168,169,172,173,176,177,180,181,184,185,188, 189,194, 195,198,199,202, 203,206,207,210,211,214,215, 218,219,222,223,224,225,228,229,232,233,236,237,240, 241,244, 245, 248, 249,252,253
s_0	1,3,5,7,9,11,13,15,16,18,20,22,24,26,28,30,33,35,37,39,41, 43,45,47,48,50,52,54,56,58,60,62, 65,67,69,71,73,75,77,79, 80,82,84,86,88,90,92,94,97,99,101,103,105,107,109,111, 112,114,116,118,120,122,124,126,129,131,133,135,137, 139,141,143,144,146,148,150,152,154,156,158,161,163, 165,167,169,171,173,175,176,178,180,182,184,186,188, 190,193,195, 197,199,201,203,205,207,208,210,212,214, 216,218,220,222,225,227,229,231,233,235,237,239,240, 242,244,246,248,250, 252,254

Fig. 4.1 Block diagram
of a nibble-size minimum-
delay complex binary
adder [1, 3]

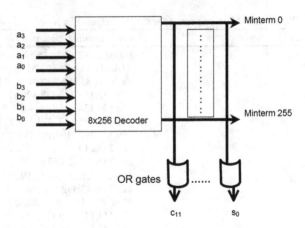

4.1.2 Ripple-Carry Adder

A nibble-size ripple-carry adder consists of typical half-adders, full-adders, and specially designed combinational circuit units. Basically, the adder performs the addition of two 4-bits complex binary numbers (augend: $a_3a_2a_1a_0$ and addend: $b_3b_2b_1b_0$) and generates a 4-bits Sum and up to 8 Extended-Carries. The block diagram of the ripple-carry adder consists of the Addition Unit, the Zero Detection Unit, the Extended-Carry Generation Unit, and the Output Generation Unit (Fig. 4.2).

The Addition Unit is structured from 4 semi-cascaded stages. Each stage is responsible for generating one of the Sum bits $(S_0S_1S_2S_3)$. In CBNS, the addition of two bits at stage i produces two carries that propagate to stages $i + 2$ and $i + 3$. Since no carry-in(s) to the adder is assumed, stages 0 and 1 are easily implemented using 2 half-adders.

Stage 2 is implemented using a full-adder with a carry-in generated from stage 0. For stage 3, a specially designed 4-input binary variables adding component has been implemented. Stage 3 performs the addition of bits a_3 and b_3 of the Augend and Addend with two possible carries referred to as K_{31} and K_{32}, which may be generated from stages 0 and 1, respectively. The stage produces result bit S_3 and two carry bits, C_3 and Q_3, according to the truth table for addition. C_3 is a normal carry due to adding three ones $(1 + 1 + 1)$, and Q_3 is an extended carry due to adding four ones $(1 + 1 + 1 + 1)$ in complex binary representation. C_3 should propagate to stages 5 and 6, and Q_3 to stages 7, 9,10, and 11. Since the adder performs 4-bit complex binary addition, the carries C_2, C_3, and Q_3 are taken to the inputs of the Extended-Carry Generation Unit in order to generate all the necessary carries. All carries generated by stages 2 and 3 are handled by dummy stages in the Extended-Carry Generation Unit, referred to by stages 4–11. This Unit would generate the extended carries $(C_4C_5C_6C_7C_8C_9C_{10}C_{11})$ as inputs to the Output Generation Unit.

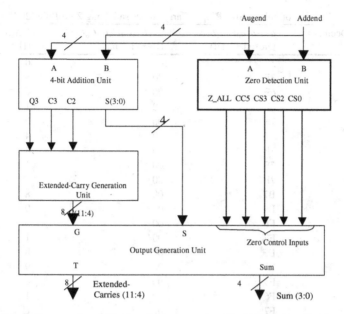

Fig. 4.2 Block diagram of a nibble-size ripple-carry complex binary adder [2, 3]

The Boolean expressions for stages 0, 1, and 2 are obvious from the use of half-adder and full-adder designs. For stage 3, the Boolean expressions for the outputs are found from the minimization of four-variable Karnaugh maps. These are:

$$S_3 = a_3 \oplus b_3 \oplus K_{31} \oplus K_{32} \tag{4.1}$$

$$C_3 = \overline{a_3}K_{31}K_{32} + b_3K_{31}\overline{K_{32}} + a_3\overline{b_3}K_{31} + a_3\overline{K_{31}}K_{32} + a_3b_3\overline{K_{31}} + b_3\overline{K_{31}}K_{32} \tag{4.2}$$

$$Q_3 = a_3b_3K_{31}K_{32} \tag{4.3}$$

The S_3 expression is a four-input odd function which can be implemented using Exclusive-OR gates, the Q_3 expression is a four-input AND function, and the C_3 expression is a sum-of-products expression which can be implemented using AND-OR or NAND-NAND logic gates.

The Zero Detection Unit determines the conditions necessary to generate special output results based on the recognition of specific patterns for the Addend and the Augend. The conditions considered are based on the Zero Rule for the complex binary addition. This Unit receives nibble-size augend and addend inputs and generates five control signals: CS_0, CS_2, CS_3, CS_5, and Z_ALL based on the

Table 4.11 Minterms of Nibble-Size Ripple-Carry Adder satisfying Zero-Rule [2, 3]

Minterm (Decimal)	$a_3a_2a_1a_0b_3b_2b_1b_0$ (Hexadecimal)	$C_{11}C_{10}C_9C_8C_7C_6C_5C_4$ (Hexadecimal)	$S_3S_2S_1S_0$ (Hexadecimal)
55	37	00	0
63	3F	00	8
110	6E	00	0
111	6F	00	1
115	73	00	0
119	77	00	4
123	7B	00	4
126	7E	00	1
127	7F	00	C
183	B7	00	8
191	BF	06	0
230	E6	00	0
231	E7	00	1
238	EE	00	8
239	EF	00	9
243	F3	00	8
246	F6	00	1
247	F7	00	C
251	FB	06	0
254	FE	00	9
255	FF	06	4

patterns satisfying the Zero Rule. Table 4.11 lists all the minterms that will generate special output results. The Boolean expressions characterizing each control input are given as follows:

$$CS_0 = \sum(111, 126, 231, 239, 246, 254) \tag{4.4}$$

$$CS_0 = (a_2a_1b_2b_1)(\overline{a_0}b_3b_0 + a_0b_3\overline{b_0} + a_3b_3(a_0 \oplus b_0)) \tag{4.5}$$

$$CS_2 = \sum(119, 127, 247, 255) = a_2a_1a_0b_2b_1b_0 \tag{4.6}$$

$$CS_3 = \sum(63, 123, 127, 183, 238, 239, 243, 247, 254) \tag{4.7}$$

$$CS_3 = (a_1b_1)(\overline{a_3}a_0b_3b_2b_0 + a_3a_0\overline{b_3}b_2b_0 + a_3a_2b_3b_2\overline{b_0} + \overline{a_3}a_2a_0b_3b_0$$
$$+a_3a_2a_0\overline{b_3}b_0 + a_3a_2\overline{a_0}b_3b_2) \tag{4.8}$$

$$CS_5 = \sum(191, 251, 255) \tag{4.9}$$

$$CS_5 = (a_3a_2a_1a_0b_3b_1b_0 + a_3a_1a_0b_3b_2b_1b_0) \tag{4.10}$$

$$Z\_ALL = \sum(155, 110, 115, 230) \tag{4.11}$$

$$Z\_ALL = (a_1b_1)(\overline{a_3}a_0\overline{b_3}b_0(a_2 \oplus b_2) + a_2\overline{a_0}b_2\overline{b_0}(a_3 \oplus b_3)) \tag{4.12}$$

The Output Generation Unit receives the control signals, (CS_0, CS_2, CS_3, CS_5, and Z_ALL), from the Zero Detection Unit, the Sum bits ($S_0S_1S_2S_3$) from the Addition Unit, and the extended-carries ($C_4C_5C_6C_7C_8C_9C_{10}C_{11}$) from the Extended Carry-Generation Unit. It, then, determines the actual result of addition (Sum (3:0) and Extended Carries (T11:T4)) according to the control signals described above.

4.1.3 State-Machine Adder

The design of this adder is based on using a state machine to store the logic details rather than designing the addition and carry operations with discrete components. This approach results in a very simple circuit implementation and does not impose any limit on the size of the operands. The entire adder consists of a few gates to add single bits from the input numbers, memory to hold the state and output information, and a register to store the current state (in effect, the carry to the next addition). Since the operations are done bit-by-bit, the adder itself imposes no limitations on the sizes of the numbers to be added. The state machine is not aware of the number of bits in the input numbers. The only requirement is to make sure that the inputs are sufficiently padded with high-order zeroes to allow for the carry from the addition of the high-order bits of the input numbers. Depending on the carry-in from the previous addition and the values of the two current bits to be added, a carry of up to 8 bits may result. Consequently, 8 bits of padding of high-order zeroes would be required to correctly complete the addition.

The logic of the adder is stored in a state table. Each entry of the table contains the next state of the state machine and the output from the last addition operation. The table is organized into three sections, one for each state transition. A transition is determined by the result of the addition of the next two input bits. There are three results—(i) 0 + 0, (ii) 0 + 1 or 1 + 0, and (iii) 1 + 1. The result selects the particular section of the state table to use for determining the next state and output. The input to the state machine is the sum of the current two binary bits to be added. The current state is composed of the carry out of the previous operation. The next state (and single bit output for the current addition) is found in the memory location formed by the concatenation of the input and current state bits as described above.

The state table is shown in Tables 4.12, 4.13 and the state diagram is given in Fig. 4.3. The state table was constructed by:

(i) Starting with a sum of 0 and a carry in (current state) of 0.
(ii) Adding 0, 1, or 2. (The sum of input bits A and B).
(iii) Shifting out the low order sum bit (The sum will be a single binary bit, 0 or 1 and a carry. The result of 0 plus 0 is 0 with no carry, 0 plus 1 is 1 with no carry, and 1 plus 1 is 0 with a carry of 110).
(iv) Repeating the above until all possible results were produced.

Table 4.12 Adder state table [4]

(1) Memory location (current state)	(2) Contents (Hex)	(3) Previous carry	(4) Add	(5) Result	(6) After shift	(7) Next state	(8) Output
00	00	0000	0000	0000	000	00	0
	80		0001	0001	000	00	1
	01		1100	1100	110	01	0
01	02	0110	0000	0110	011	02	0
	82		0001	0111	011	02	1
	05		1100	111010	11101	05	0
02	83	0011	0000	0011	001	03	1
	04		0001	1110	111	04	0
	84		1100	1111	111	04	1
03	80	0001	0000	0001	000	00	1
	01		0001	1100	110	01	0
	81		1100	1101	110	01	1
04	82	0111	0000	0111	011	02	1
	05		0001	111010	11101	05	0
	85		1100	111011	11101	05	1
05	86	11101	0000	11101	1110	06	1
	00		0001	0000	000	00	0
	80		1100	0001	000	00	1
06	04	1110	0000	1110	111	04	0
	84		0001	1111	111	04	1
	07		1100	111010010	11101001	07	0
07	88	11101001	0000	11101001	1110100	08	1
	09		0001	0100	010	09	0
	89		1100	0101	010	09	1
08	0A	1110100	0000	1110100	111010	10	0
	8A		0001	1110101	111010	10	1
	0B		1100	1000	100	11	0
09	03	0010	0000	0010	001	03	0
	83		0001	0011	001	03	1
	04		1100	1110	111	04	0
10	05	111010	0000	111010	11101	05	0
	85		0001	111011	11101	05	1
	0D		1100	111010110	11101011	13	0
11	0C	100	0000	0100	010	12	0
	8C		0001	0101	010	12	1
	0E		1100	111000	11100	14	0
12	03	0010	0000	0010	001	03	0
	83		0001	0011	001	03	1
	04		1100	1110	111	04	0

It can be seen that there are 15 states (representing carries from a previous operation) and 3 inputs per state resulting in 45 state transitions. For each transition the output is 1 or 0. For the purposes of this implementation, an additional redundant state was added simply to fill the 16 memory locations that were

Table 4.13 Adder state table [4]

(1) Memory location (current state)	(2) Contents (hex)	(3) Previous carry	(4) Add	(5) Result	(6) After shift	(7) Next state	(8) Output
13	8F	11101011	0000	11101011	1110101	15	1
	02		0001	0110	011	02	0
	82		1100	0111	011	02	1
14	06	11100	0000	11100	1110	06	0
	86		0001	11101	1110	06	1
	00		1100	0000	000	00	0
15	8A	1110101	0000	1110101	111010	10	1
	0B		0001	1000	100	11	0
	8B		1100	1001	100	11	1

available. This additional state serves no other purpose and could be deleted with no effect on the operation of the adder.

It should be noted that the adder does not actually add inputs A and B to a previous carry operation. It only changes state based on the current input and current state. The states represent the result of addition operations. The actual result (0 or 1) of the addition of the two current input bits is stored in the state memory and is generated as output as the state changes.

Following is a description of the columns of the adder state table (Tables 4.12, 4.13):

Column Number (1) Memory Location (Current State): Memory is arranged in three banks. Each bank contains 16 locations, 00–15 representing the current state. In a physical implementation various combinations of memory sizes could be used. For example a single 48-location memory or three 16-location memories could be used. Column (1) is the memory location number in decimal. The location is shown only for the first bank of memory. There are three rows of data for each location. Each row corresponds to a bank of memory. For example for location '00', the first row corresponds to location '00' for bank 1, the second row to location '00' for bank 2, and the third row to location '00' for bank 3.

Column Number (2) Contents (Hex): The contents of each memory location of each bank are shown in hexadecimal. For example, for memory location '06', bank 1 contains x'04' or binary 00000100. Bank 2 contains x'84' or binary 10000100. These values represent the output and next state. The output bit is in the high order location (leftmost bit). The next state is in the 4 low order bits (4 rightmost bits). For example, memory location 13, bank1 contains x'8F' or binary 10001111. The high order bit is 1 and is the output bit when the current state is 13 and 0000 (see column 3 and 4 explanation) is added via input bits A and B (both A and B are 0). The four low order bits '1111' are the next state. The next state is shown in decimal in column 7. It corresponds to the four low order bits of column 2.

Column Number (3) Previous Carry: This column shows the carry from the result of the addition of the previous two bits. The memory location in column 1 is

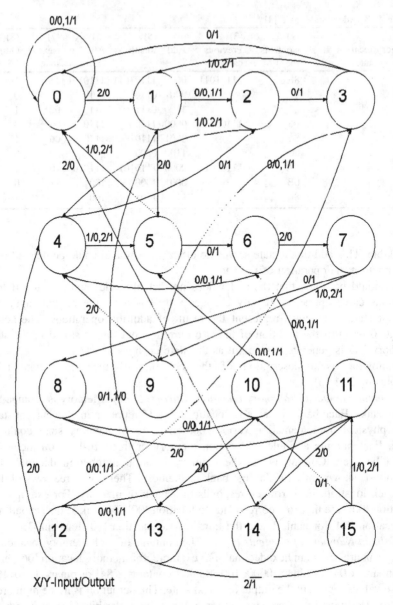

Fig. 4.3 Adder state diagram [4]

the representation of this value. The addition of input numbers A and B and the previous carry is done in the adder by considering the previous carry as the current state and the sum of A and B as selecting the particular state transition from the current state.

Column Number (4) Add: This column shows the sum of input bits A and B that is added to the previous carry. The sum actually results in a selection of the memory bank to use for finding the next state. If the sum is 0000 (both A and B are 0), bank 1 is used. If the sum is 0001 (either A or B is one), bank 2 is used. If the sum is 1100 (both A and B are 1), bank 3 is used. The value of the previous state is used to access the memory location of the selected bank to get the next state and output bit.

Column Number (5) Result: The result of adding columns (4) and (3) is shown. This addition is not actually performed by the adder. This result after shifting right by one bit is assigned a state number to be used for the next operation. The result after shifting is in column (6). The state number is shown in column (7).

Column Number (6) After Shift: This column shows the result of shifting column 5 right by one bit. This is the operation that would be performed to get ready to add the next two values of input bits A and B. The value is assigned a state number (column 7) that is the next state (column 1).

Column Number (7) Next State: This number represents the value in column (6). It is the memory location that is used in the next decoding operation. The sum of input bits A and B select the memory bank to use. Column (6) addresses the location within the bank.

Column Number (8) Output: This is the bit that is output as a result of adding A, B, and the carry operation from the previous addition operation.

Let us add 1 and 1 using the state diagram (Fig. 4.3). Assuming the initial state is 0, the addition of input bit A $= 1$ and B $= 1$ results in an input of 2 (in decimal), output of 0, and next state of 1. The next values of A and B are 0, so the next input is 0, output is 0 and next state is 2. Again A and B are 0, the input is 0, the output now is 1, and the next state is 3. Again A and B are 0, the input is 0, the output is again 1, and the next state is 0. (The output to this point is $1100_{Base\ (-1+j)}$ which is 2_{10} as it should be). A and B are 0 for the rest of the 32-bit sequence. The input is 0, the output is 0, and the next state is 0 for all remaining input bits. The same sequence can be followed in the state table.

4.1.4 Implementations and Performance Evaluations

Both minimum-delay and ripple-carry adders (nibble-size) have been implemented on Xilinx Virtex FPGA and implementation statistics obtained are given in Table 4.14 [3].

Traditional Base-2 binary adder implementation statistics are also given in Table 4.14. The speed-up comparisons of the four adders' designs are given in Table 4.15.

A functional diagram of the state-machine adder, which can be implemented, is shown in Fig. 4.4 [4]. Data are input by storing two complex binary base numbers in the input memories. The numbers are shifted serially, least significant bit first

Table 4.14 Implementation statistics for nibble-size adder designs on Xilinx Virtex FPGA [3]

	Complex binary adders		Base-2 binary adders	
	Minimum-delay	Ripple-carry	Minimum-delay	Ripple-carry
Number of external IOBs	20/94 (21 %)	20/94 (21 %)	13/94 (13 %)	13/94 (13 %)
Number of slices	455/768 (59 %)	31/768 (4 %)	391/768 (50 %)	6/768 (1 %)
Number of 4 input LUTs	857/1536 (55 %)	59/1536 (3 %)	755/1536 (49 %)	9/1536 (1 %)
Number of bonded IOBs	20/94 (21 %)	20/94 (21 %)	13/94 (13 %)	13/94 (13 %)
Gate count	5142	354	4530	54
Average connection delay (ns)	3.179	1.640	3.169	1.525
Maximum pin delay (ns)	11.170	4.024	9.207	2.421
Maximum combinational delay (ns)	32.471	24.839	28.442	15.389
Score of the design	450	218	443	188

Table 4.15 Speed-up comparisons of adders' designs [3]

	Minimum-delay adders		Ripple-carry adders	
	Base $(-1 + j)$	Base 2	Base $(-1 + j)$	Base 2
Minterms	256	256	256	256
Number of minterms giving different outputs (M)	175/256 (68 %)	175/256 (68 %)	175/256 (68 %)	175/256 (68 %)
Number of minterms giving same outputs (m)	81/256 (32 %)	81/256 (32 %)	81/256 (32 %)	81/256 (32 %)
Time taken to compute all M-terms (2:1)	5682.425 ns	9954.70 ns	4346.825 ns	5386.15 ns
Time taken to compute all m-terms (1:1)	2630.151 ns	2303.802 ns	2011.959 ns	1246.509 ns
Total time taken to compute all M + m-terms	8312.576 ns	12258.502 ns	6358.784 ns	6632.659 ns
Average time for adding two nibble-size operands	32.471 ns	47.885 ns	24.839 ns	25.909 ns
Speed-up	REF + 32 %	REF	ref + 4 %	ref

into the single-bit adder section. As discussed previously the adder merely selects a memory bank to use based on the values of the input bits (00, 01 or 10, or 11). The 'state and output' memory holds the state values and output for a given state transition. The result is shifted into the output shift registers and the carry is saved in a register for addition to the next two input bits. A master clock synchronizes all operations. An implementation (summarized below) allows addition of two 32-bit

Fig. 4.4 State-machine adder functional diagram [4]

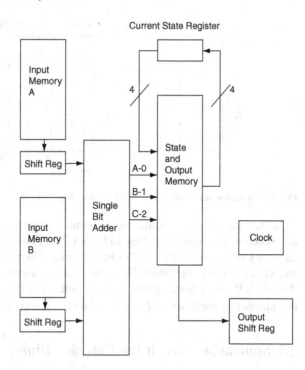

numbers. No provision was made for carries beyond 32 bits. In the earlier discussion of the adder state machine it was noted that a carry out value could be up to 8 bits. So, if the input numbers are limited to 24 bits, a maximum size carry could always be handled.

Addition of numbers of any length could be accommodated by expanding the input memory and output shift register length. The adder itself is a single-bit adder. An actual implementation in a digital computing system would use some other form of input and output such as general-purpose registers. Also, provision for carry beyond the maximum register length would be made. In short, there is no limitation to the length of numbers that can be added other than that imposed by the input and output devices. An implementation of the adder was accomplished using System-View, a modeling and simulation tool that provides tokens representing memories, gates, and other devices from which systems may be constructed. The results of output operations can be plotted using 'sink' tokens. For example the display of the results of an addition shifted into the output shift register is shown in Fig. 4.5. On the plot of Token 57 data, the results are read right to left (least significant bit of the sum is on the left). The timescale is arbitrary.

In a physical implementation of the circuit, the main component of delay and thus a limit on performance would typically be the time required to address the memory with the results of the addition operation and store the result in the 'current state' register. Otherwise the addition speed would be limited only by the speed of the logic gates at input and output.

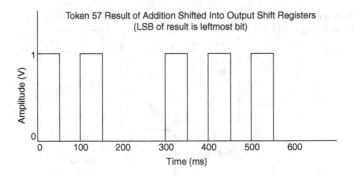

Fig. 4.5 Addition result shifted into output shift registers [4]

In the SystemView implementation numbers to be added (for example 2 and $-j$) are placed in the input memories. In CBNS, 2 is 1100 or xC in hexadecimal. Similarly, in CBNS, $-j$ is 111 or $x7$ in hexadecimal. So, xC is entered in one memory and $x7$ is entered in the other. The result can be viewed from the plot of Token 57 (Fig. 4.5). Reading from right to left the result is $111011_{\text{Base }(-1+j)}$. An expansion of this number as coefficients of powers of $(-1+j)$ shows it to be $(2-j)$.

4.2 Subtractor Circuit for Complex Binary Numbers

A minimum-delay nibble-size subtractor for complex binary numbers has been presented in [5, 6].

4.2.1 Minimum-Delay Subtractor

The minimum-delay nibble-size CBNS subtractor has been designed following a procedure similar to what has been described for the adder circuit in Chap. 3. The procedure involves writing a truth table with four-bit minuend ($a_3a_2a_1a_0$) and subtrahend ($b_3b_2b_1b_0$) operands as inputs (total of $2^8 = 256$ minterms) and 11 outputs ($d_{10}d_9d_8d_7d_6d_5d_4d_3d_2d_1d_0$) which are obtained by subtracting each pair of nibble-size inputs according to the subtraction procedure described in Chap. 3. Each output is then expressed in sum-of-minterms form. The resulting design expressions have been implemented using an 8×256 decoder (to generate each minterm) and multiple-input OR gates (to combine relevant minterms for each output).

Tables 4.16, 4.17, 4.18, 4.19, 4.20, 4.21, 4.22, 4.23 present complete truth table for a nibble-size minimum-delay complex binary subtractor. For the sake of simplicity the 11 outputs have been collectively labeled as "Difference" in these tables. The sum-of-minterms expressions for outputs of the subtractor are listed in Tables 4.24, 4.25. Block diagram of the subtractor is given in Fig. 4.6.

Table 4.16 Truth table for a Nibble-size minimum-delay complex binary subtractor [5, 6] (Minterm: $a_3a_2a_1a_0$ SUB $b_3b_2b_1b_0 = d_{10}d_9d_8d_7d_6d_5d_4d_3d_2d_1d_0$)

Minterm	Minuend				Subtrahend				Difference
	a_3	a_2	a_1	a_0	b_3	b_2	b_1	b_0	$d_{10}d_9...d_1d_0$
0	0	0	0	0	0	0	0	0	00000000000
1	0	0	0	0	0	0	0	1	00000011101
2	0	0	0	0	0	0	1	0	00000111010
3	0	0	0	0	0	0	1	1	00000000111
4	0	0	0	0	0	1	0	0	00001110100
5	0	0	0	0	0	1	0	1	00000011001
6	0	0	0	0	0	1	1	0	00000001110
7	0	0	0	0	0	1	1	1	00000000011
8	0	0	0	0	1	0	0	0	00011101000
9	0	0	0	0	1	0	0	1	00000010101
10	0	0	0	0	1	0	1	0	00000110010
11	0	0	0	0	1	0	1	1	00011101111
12	0	0	0	0	1	1	0	0	00000011100
13	0	0	0	0	1	1	0	1	00000010001
14	0	0	0	0	1	1	1	0	00000000110
15	0	0	0	0	1	1	1	1	00011101011
16	0	0	0	1	0	0	0	0	00000000001
17	0	0	0	1	0	0	0	1	00000000000
18	0	0	0	1	0	0	1	0	00000111011
19	0	0	0	1	0	0	1	1	00000111010
20	0	0	0	1	0	1	0	0	00001110101
21	0	0	0	1	0	1	0	1	00001110100
22	0	0	0	1	0	1	1	0	00000001111
23	0	0	0	1	0	1	1	1	00000001110
24	0	0	0	1	1	0	0	0	00011101001
25	0	0	0	1	1	0	0	1	00011101000
26	0	0	0	1	1	0	1	0	00000110011
27	0	0	0	1	1	0	1	1	00000110010
28	0	0	0	1	1	1	0	0	00000011101
29	0	0	0	1	1	1	0	1	00000011100
30	0	0	0	1	1	1	1	0	00000000111
31	0	0	0	1	1	1	1	1	00000000110

Table 4.17 Truth table for a Nibble-size minimum-delay complex binary subtractor [5, 6] (Minterm: $a_3a_2a_1a_0$ SUB $b_3b_2b_1b_0 = d_{10}d_9d_8d_7d_6d_5d_4d_3d_2d_1d_0$)

Minterm	Minuend				Subtrahend				Difference
	a_3	a_2	a_1	a_0	b_3	b_2	b_1	b_0	$d_{10}d_9...d_1d_0$
32	0	0	1	0	0	0	0	0	00000000010
33	0	0	1	0	0	0	0	1	00000011111
34	0	0	1	0	0	0	1	0	00000000000
35	0	0	1	0	0	0	1	1	00000011101
36	0	0	1	0	0	1	0	0	00001110110
37	0	0	1	0	0	1	0	1	00000011011
38	0	0	1	0	0	1	1	0	00001110100
39	0	0	1	0	0	1	1	1	00000011001
40	0	0	1	0	1	0	0	0	00011101010
41	0	0	1	0	1	0	0	1	00000010111
42	0	0	1	0	1	0	1	0	00011101000
43	0	0	1	0	1	0	1	1	00000010101
44	0	0	1	0	1	1	0	0	00000011110
45	0	0	1	0	1	1	0	1	00000010011
46	0	0	1	0	1	1	1	0	00000011100
47	0	0	1	0	1	1	1	1	00000010001
48	0	0	1	1	0	0	0	0	00000000011
49	0	0	1	1	0	0	0	1	00000000010
50	0	0	1	1	0	0	1	0	00000000001
51	0	0	1	1	0	0	1	1	00000000000
52	0	0	1	1	0	1	0	0	00001110111
53	0	0	1	1	0	1	0	1	00001110110
54	0	0	1	1	0	1	1	0	00001110101
55	0	0	1	1	0	1	1	1	00001110100
56	0	0	1	1	1	0	0	0	00011101011
57	0	0	1	1	1	0	0	1	00011101010
58	0	0	1	1	1	0	1	0	00011101001
59	0	0	1	1	1	0	1	1	00011101000
60	0	0	1	1	1	1	0	0	00000011111
61	0	0	1	1	1	1	0	1	00000011110
62	0	0	1	1	1	1	1	0	00000011101
63	0	0	1	1	1	1	1	1	00000011100

Table 4.18 Truth table for a Nibble-size minimum-delay complex binary subtractor [5, 6] (Minterm: $a_3a_2a_1a_0$ SUB $b_3b_2b_1b_0 = d_{10}d_9d_8d_7d_6d_5d_4d_3d_2d_1d_0$)

Minterm	Minuend				Subtrahend				Difference
	a_3	a_2	a_1	a_0	b_3	b_2	b_1	b_0	$d_{10}d_9...d_1d_0$
64	0	1	0	0	0	0	0	0	00000000100
65	0	1	0	0	0	0	0	1	00011101001
66	0	1	0	0	0	0	1	0	00000111110
67	0	1	0	0	0	0	1	1	00000110011
68	0	1	0	0	0	1	0	0	00000000000
69	0	1	0	0	0	1	0	1	00000000111
70	0	1	0	0	0	1	1	0	00000111010
71	0	1	0	0	0	1	1	1	00000000111
72	0	1	0	0	1	0	0	0	00011101100
73	0	1	0	0	1	0	0	1	00011100001
74	0	1	0	0	1	0	1	0	00000110110
75	0	1	0	0	1	0	1	1	11101011011
76	0	1	0	0	1	1	0	0	00011101000
77	0	1	0	0	1	1	0	1	00000010101
78	0	1	0	0	1	1	1	0	00000110010
79	0	1	0	0	1	1	1	1	00011101111
80	0	1	0	1	0	0	0	0	00000000101
81	0	1	0	1	0	0	0	1	00000000100
82	0	1	0	1	0	0	1	0	00000111111
83	0	1	0	1	0	0	1	1	00000111110
84	0	1	0	1	0	1	0	0	00000000001
85	0	1	0	1	0	1	0	1	00000000000
86	0	1	0	1	0	1	1	0	00000111011
87	0	1	0	1	0	1	1	1	00000111010
88	0	1	0	1	1	0	0	0	00011101101
89	0	1	0	1	1	0	0	1	00011101100
90	0	1	0	1	1	0	1	0	00000110111
91	0	1	0	1	1	0	1	1	00000110110
92	0	1	0	1	1	1	0	0	00011101001
93	0	1	0	1	1	1	0	1	00011101000
94	0	1	0	1	1	1	1	0	00000110011
95	0	1	0	1	1	1	1	1	00000110010

Table 4.19 Truth table for a Nibble-size minimum-delay complex binary subtractor [5, 6] (Minterm: $a_3a_2a_1a_0$ SUB $b_3b_2b_1b_0 = d_{10}d_9d_8d_7d_6d_5d_4d_3d_2d_1d_0$)

Minterm	Minuend				Subtrahend				Difference
	a_3	a_2	a_1	a_0	b_3	b_2	b_1	b_0	$d_{10}d_9...d_1d_0$
96	0	1	1	0	0	0	0	0	00000000110
97	0	1	1	0	0	0	0	1	00011101011
98	0	1	1	0	0	0	1	0	00000000100
99	0	1	1	0	0	0	1	1	00011101001
100	0	1	1	0	0	1	0	0	00000000010
101	0	1	1	0	0	1	0	1	00000011111
102	0	1	1	0	0	1	1	0	00000000000
103	0	1	1	0	0	1	1	1	00000011101
104	0	1	1	0	1	0	0	0	00011101110
105	0	1	1	0	1	0	0	1	00011100011
106	0	1	1	0	1	0	1	0	00011101100
107	0	1	1	0	1	0	1	1	00011100001
108	0	1	1	0	1	1	0	0	00011101010
109	0	1	1	0	1	1	0	1	00000010111
110	0	1	1	0	1	1	1	0	00011101000
111	0	1	1	0	1	1	1	1	00000010101
112	0	1	1	1	0	0	0	0	00000000111
113	0	1	1	1	0	0	0	1	00000000110
114	0	1	1	1	0	0	1	0	00000000101
115	0	1	1	1	0	0	1	1	00000000100
116	0	1	1	1	0	1	0	0	00000000011
117	0	1	1	1	0	1	0	1	00000000010
118	0	1	1	1	0	1	1	0	00000000001
119	0	1	1	1	0	1	1	1	00000000000
120	0	1	1	1	1	0	0	0	00011101111
121	0	1	1	1	1	0	0	1	00011101110
122	0	1	1	1	1	0	1	0	00011101101
123	0	1	1	1	1	0	1	1	00011101100
124	0	1	1	1	1	1	0	0	00011101011
125	0	1	1	1	1	1	0	1	00011101010
126	0	1	1	1	1	1	1	0	00011101001
127	0	1	1	1	1	1	1	1	00011101000

Table 4.20 Truth table for a Nibble-size minimum-delay complex binary subtractor [5, 8] (Minterm: $a_3a_2a_1a_0$ SUB $b_3b_2b_1b_0 = d_{10}d_9d_8d_7d_6d_5d_4d_3d_2d_1d_0$)

Minterm	Minuend				Subtrahend				Difference
	a3	a2	a1	a0	b3	b2	b1	b0	d10d9...d1d0
128	1	0	0	0	0	0	0	0	00000001000
129	1	0	0	0	0	0	0	1	00001110101
130	1	0	0	0	0	0	1	0	00111010010
131	1	0	0	0	0	0	1	1	00000001111
132	1	0	0	0	0	1	0	0	00001111100
133	1	0	0	0	0	1	0	1	00001110001
134	1	0	0	0	0	1	1	0	00001100110
135	1	0	0	0	0	1	1	1	00000001011
136	1	0	0	0	1	0	0	0	00000000000
137	1	0	0	0	1	0	0	1	00000011101
138	1	0	0	0	1	0	1	0	00000000110
139	1	0	0	0	1	0	1	1	00000000111
140	1	0	0	0	1	1	0	0	00001110100
141	1	0	0	0	1	1	0	1	00000011001
142	1	0	0	0	1	1	1	0	00000001110
143	1	0	0	0	1	1	1	1	00000000011
144	1	0	0	1	0	0	0	0	00000001001
145	1	0	0	1	0	0	0	1	00000001000
146	1	0	0	1	0	0	1	0	00111010011
147	1	0	0	1	0	0	1	1	00111010010
148	1	0	0	1	0	1	0	0	00001111101
149	1	0	0	1	0	1	0	1	00001111100
150	1	0	0	1	0	1	1	0	00001100111
151	1	0	0	1	0	1	1	1	00001100110
152	1	0	0	1	1	0	0	0	00000000001
153	1	0	0	1	1	0	0	1	00000000000
154	1	0	0	1	1	0	1	0	00000111011
155	1	0	0	1	1	0	1	1	00000111010
156	1	0	0	1	1	1	0	0	00001110101
157	1	0	0	1	1	1	0	1	00001110100
158	1	0	0	1	1	1	1	0	00000001111
159	1	0	0	1	1	1	1	1	00000001110

Table 4.21 Truth table for a Nibble-size minimum-delay complex binary subtractor [5, 8] (Minterm: $a_3a_2a_1a_0$ SUB $b_3b_2b_1b_0 = d_{10}d_9d_8d_7d_6d_5d_4d_3d_2d_1d_0$)

Minterm	Minuend				Subtrahend				Difference
	a_3	a_2	a_1	a_0	b_3	b_2	b_1	b_0	$d_{10}d_9...d_1d_0$
160	1	0	1	0	0	0	0	0	00000001010
161	1	0	1	0	0	0	0	1	00001110111
162	1	0	1	0	0	0	1	0	00000001000
163	1	0	1	0	0	0	1	1	00001110101
164	1	0	1	0	0	1	0	0	00001111110
165	1	0	1	0	0	1	0	1	00001110011
166	1	0	1	0	0	1	1	0	00001111100
167	1	0	1	0	0	1	1	1	00001110001
168	1	0	1	0	1	0	0	0	00000000010
169	1	0	1	0	1	0	0	1	00000011111
170	1	0	1	0	1	0	1	0	00000000000
171	1	0	1	0	1	0	1	1	00000011101
172	1	0	1	0	1	1	0	0	00001110110
173	1	0	1	0	1	1	0	1	00000011011
174	1	0	1	0	1	1	1	0	00001110100
175	1	0	1	0	1	1	1	1	00000011001
176	1	0	1	1	0	0	0	0	00000001011
177	1	0	1	1	0	0	0	1	00000001010
178	1	0	1	1	0	0	1	0	00000001001
179	1	0	1	1	0	0	1	1	00000001000
180	1	0	1	1	0	1	0	0	00001111111
181	1	0	1	1	0	1	0	1	00001111110
182	1	0	1	1	0	1	1	0	00001111101
183	1	0	1	1	0	1	1	1	00001111100
184	1	0	1	1	1	0	0	0	00000000011
185	1	0	1	1	1	0	0	1	00000000010
186	1	0	1	1	1	0	1	0	00000000001
187	1	0	1	1	1	0	1	1	00000000000
188	1	0	1	1	1	1	0	0	00001110111
189	1	0	1	1	1	1	0	1	00001110110
190	1	0	1	1	1	1	1	0	00001110101
191	1	0	1	1	1	1	1	1	00001110100

Table 4.22 Truth table for a Nibble-size minimum-delay complex binary subtractor [5, 8] (Minterm: $a_3a_2a_1a_0$ SUB $b_3b_2b_1b_0 = d_{10}d_9d_8d_7d_6d_5d_4d_3d_2d_1d_0$)

Minterm	Minuend				Subtrahend				Difference
	a_3	a_2	a_1	a_0	b_3	b_2	b_1	b_0	$d_{10}d_9...d_1d_0$
192	1	1	0	0	0	0	0	0	00000001100
193	1	1	0	0	0	0	0	1	00000000001
194	1	1	0	0	0	0	1	0	00111010110
195	1	1	0	0	0	0	1	1	00000111011
196	1	1	0	0	0	1	0	0	00000001000
197	1	1	0	0	0	1	0	1	00001110101
198	1	1	0	0	0	1	1	0	00111010010
199	1	1	0	0	0	1	1	1	00000001111
200	1	1	0	0	1	0	0	0	00000000100
201	1	1	0	0	1	0	0	1	00011101001
202	1	1	0	0	1	0	1	0	00000111110
203	1	1	0	0	1	0	1	1	00000110011
204	1	1	0	0	1	1	0	0	00000000000
205	1	1	0	0	1	1	0	1	00000011101
206	1	1	0	0	1	1	1	0	00000111010
207	1	1	0	0	1	1	1	1	00000000111
208	1	1	0	1	0	0	0	0	00000001101
209	1	1	0	1	0	0	0	1	00000001100
210	1	1	0	1	0	0	1	0	00111010111
211	1	1	0	1	0	0	1	1	00111010110
212	1	1	0	1	0	1	0	0	00000001001
213	1	1	0	1	0	1	0	1	00000001000
214	1	1	0	1	0	1	1	0	00111010011
215	1	1	0	1	0	1	1	1	00111010010
216	1	1	0	1	1	0	0	0	00000000101
217	1	1	0	1	1	0	0	1	00000000100
218	1	1	0	1	1	0	1	0	00000111111
219	1	1	0	1	1	0	1	1	00000111110
220	1	1	0	1	1	1	0	0	00000000001
221	1	1	0	1	1	1	0	1	00000000000
222	1	1	0	1	1	1	1	0	00000111011
223	1	1	0	1	1	1	1	1	00000111010

Table 4.23 Truth table for a Nibble-size minimum-delay complex binary subtractor [5, 8] (Minterm: $a_3a_2a_1a_0$ SUB $b_3b_2b_1b_0 = d_{10}d_9d_8d_7d_6d_5d_4d_3d_2d_1d_0$)

Minterm	Minuend				Subtrahend				Difference
	a_3	a_2	a_1	a_0	b_3	b_2	b_1	b_0	$d_{10}d_9...d_1d_0$
224	1	1	1	0	0	0	0	0	00000001110
225	1	1	1	0	0	0	0	1	00000000011
226	1	1	1	0	0	0	1	0	00000001100
227	1	1	1	0	0	0	1	1	00000000001
228	1	1	1	0	0	1	0	0	00000001010
229	1	1	1	0	0	1	0	1	00001110111
230	1	1	1	0	0	1	1	0	00000001000
231	1	1	1	0	0	1	1	1	00001110101
232	1	1	1	0	1	0	0	0	00000000110
233	1	1	1	0	1	0	0	1	00011101011
234	1	1	1	0	1	0	1	0	00000000101
235	1	1	1	0	1	0	1	1	00011101001
236	1	1	1	0	1	1	0	0	00000000010
237	1	1	1	0	1	1	0	1	00000011111
238	1	1	1	0	1	1	1	0	00000000000
239	1	1	1	0	1	1	1	1	00000011101
240	1	1	1	1	0	0	0	0	00000001111
241	1	1	1	1	0	0	0	1	00000001110
242	1	1	1	1	0	0	1	0	00000001101
243	1	1	1	1	0	0	1	1	00000001100
244	1	1	1	1	0	1	0	0	00000001011
245	1	1	1	1	0	1	0	1	00000001010
246	1	1	1	1	0	1	1	0	00000001001
247	1	1	1	1	0	1	1	1	00000001000
248	1	1	1	1	1	0	0	0	00000000111
249	1	1	1	1	1	0	0	1	00000000110
250	1	1	1	1	1	0	1	0	00000000101
251	1	1	1	1	1	0	1	1	00000000100
252	1	1	1	1	1	1	0	0	00000000011
253	1	1	1	1	1	1	0	1	00000000010
254	1	1	1	1	1	1	1	0	00000000001
255	1	1	1	1	1	1	1	1	00000000000

Table 4.24 Minterms corresponding to outputs of a Nibble-size minimum-delay subtractor [5, 8]

Subtractor Outputs	Corresponding minterms
d_{10}	75
d_9	75
d_8	75,130,146,147,194,198,210,211,214,215
d_7	8,11,15,24,25,40,42,56,57,58,59,65,72,73,76,79,88,89,92, 93,97,99,104,105,106,107,108,110,120,121,122,123,124, 125,126,127,130,146,147,194,198,201,210,211,214,215, 233,235
d_6	4,8,11,15,20,21,24,25,36,38,40,42,52,53,54,55,56,57,58,59, 65,72,73,75,76,79,88,89,93,97,99,104,105,106,107,108,110, 120,121,122,123,124,125,126,127,129,130,132,133,134, 140,146,147,148,149,150,151,156,157,161,163,164,165, 166,167,172,174,180,181,182,183,188,189,190,191,194, 197,198,201,210,211,214,215,229,231,233,235
d_5	2,4,8,10,11,15,18,19,20,21,24,25,26,27,36,38,40,42,52,53, 54,55,56,57,58,59,65,66,67,70,72,73,74,76,78,79,82,83,86, 87,88,89,90,91,92,93,94,95,97,99,104,105,106,107,108,110, 120,121,122,123,124,125,126,127,129,132,133,134,140, 148,149,150,151,154,155,156,157,161,163,164,165,166, 167,172,174,180,181,182,183,188,189,190,191,195,197, 201,202,203,206,218,219,222,223,229,231,233,235
d_4	1,2,4,5,9,10,12,13,18,19,20,21,26,27,28,29,33,35,36,37,38, 39,41,43,44,45,46,47,52,53,54,55,60,61,62,63,66,67,70,74, 75,77,78,82,83,86,87,90,91,94,95,101,103,109,111,129,130, 132,133,137,140,141,146,147,148,149,154,155,156,157, 161,163,164,165,166,167,169,171,172,173,174,175,180, 181,182,183,188,189,190,191,194,195,197,198,202,203, 205,206,210,211,214,215,218,219,222,223,229,231,237, 239
d_3	1,2,5,6,8,11,12,15,18,19,22,23,24,25,28,29,33,35,37,39,40, 42,44,46,56,57,58,59,60,61,62,63,65,66,70,72,75,76,79,82, 83,86,87,88,89,92,93,97,99,101,103,104,106,108,110,120, 121,122,123,124,125,126,127,128,131,132,135,137,141, 142,144,145,148,149,154,155,158,159,160,162,164,166, 169,171,173,175,176,177,178,179,180,181,182,183,192, 195,196,199,201,202,205,206,208,209,212,213,218,219, 222,223,224,226,228,230,233,235,237,239,240,241,242, 243,244,245,246,247
d_2	1,3,4,6,9,11,12,14,20,21,22,23,28,29,30,31,33,35,36,38,41, 43,44,46,52,53,54,55,60,61,62,63,64,66,69,71,72,74,77,79, 80,81,82,83,88,89,90,91,96,98,101,103,104,106,109,111, 112,113,114,115,120,121,122,123,129,131,132,134,137, 138,139,140,142,148,149,150,151,156,157,158,159,161, 163,164,166,169,171,172,174,180,181,182,183,188,189, 190,191,192,194,197,199,200,202,205,207,208,209,210, 211,216,217,218,219,224,226,229,231,232,234,237,239, 240,241,242,243,248,249,250,251

Table 4.25 Minterms corresponding to outputs of a Nibble-size minimum-delay subtractor [5, 8]

Subtractor Outputs	Corresponding minterms
d_1	2,3,6,7,10,11,14,15,18,19,22,23,26,27,30,31,32,33,36,37, 40,41,44,45,48,49,52,53,56,57,60,61,66,67,69,70,71,74,75,78, 79,82,83,86,87,90,91,94,95,96,97,100,101,104,105,108,109, 112,113,116,117,120,121,124,125,130,131,134,135,138, 139,142,143,146,147,150,151,154,155,158,159,160,161, 164,165,168,169,172,173,176,177,180,181,184,185,188, 189,194,195,198,199,202,203,206,207,210,211,214,215, 218,219,222,223,224,225,228,229,232,233,236,237,240, 241,244,245,248,249,252,253
d_0	1,3,5,7,9,11,13,15,16,18,20,22,24,26,28,30,33,35,37,39,41, 43,45,47,48,50,52,54,56,58,60,62,65,67,69,71,73,75,77,79, 80,82,84,86,88,90,92,94,97,99,101,103,105,107,109,111, 112,114,116,118,120,122,124,126,129,131,133,135,137, 139,141,143,144,146,148,150,152,154,156,158,161,163, 165,167,169,171,173,175,176,178,180,182,184,186,188, 190,193,195,197,199,201,203,205,207,208,210,212,214, 216,218,220,222,225,227,229,231,233,234,235,237,239, 240,242,244,246,248,250,252,254

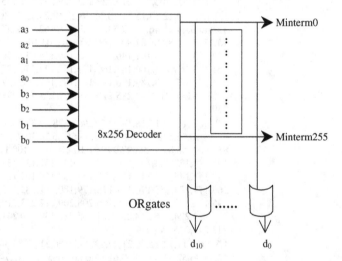

Fig. 4.6 Block diagram of a nibble-size minimum-delay complex binary subtractor [6]

4.2.2 *Implementations*

The minimum-delay subtractor has been implemented on various Xilinx FPGAs (Virtex V50CS144, Virtex2 2V10000FF1517, Spartan2 2S200PQ208, and Spartan2 2S30PQ208) and the statistics obtained are listed in Table 4.26 [6].

Table 4.26 Implementation statistics for nibble-size subtractor design on Xilinx FPGAs [6]

	Complex binary minimum-delay subtractor			
	Virtex V50CS144	Virtex2 2V10000FF1517	Spartan 2S200PQ208	Spartan 2S30PQ208
Number of external IOBs	19/94 (20 %)	19/1108 (1 %)	19/140 (13 %)	19/132 (14 %)
Number of slices	370/768 (48 %)	370/61440 (1 %)	370/2352 (15 %)	370/432 (85 %)
Number of 4 input LUTs	734/1536 (47 %)	734/122880 (1 %)	734/4704 (15 %)	734/864 (84 %)
Number of bonded IOBs	19/94 (20 %)	19/1108 (1 %)	19/140 (13 %)	19/132 (14 %)
Gate count	5040	5040	5040	5040
Maximum net delay (ns)	10.433	10.494	15.203	6.670
Maximum combinational delay (ns)	34.259	26.289	37.357	27.135

4.3 Multiplier Circuit for Complex Binary Numbers

A minimum-delay nibble-size multiplier for complex binary numbers has been presented in [5, 7].

4.3.1 Minimum-Delay Multiplier

The minimum-delay nibble-size CBNS multiplier has been designed following a procedure similar to what has been described for the adder and subtractor circuits in Sects. 4.1.1 and 4.2.1 respectively. The procedure involves writing a truth table with four-bit multiplicand ($a_3a_2a_1a_0$) and multiplier ($b_3b_2b_1b_0$) operands as inputs (total of $2^8 = 256$ minterms) and 12 outputs ($p_{11}p_{10}p_9p_8p_7p_6p_5p_4p_3p_2p_1p_0$) which are obtained by multiplying each pair of nibble-size inputs according to the multiplication procedure described in Chap. 3. Each output is then expressed in sum-of-minterms form. The resulting design expressions have been implemented using an 8 × 256 decoder (to generate each minterm) and multiple-input OR gates (to combine relevant minterms for each output).

Tables 4.27, 4.28, 4.29, 4.30, 4.31, 4.32, 4.33, 4.34 present complete truth table for a nibble-size minimum-delay complex binary multiplier circuit and the sum-of-minterms expressions for outputs of the multiplier circuit are listed in Table 4.35, 4.36. Block diagram of the multiplier circuit is given in Fig. 4.7.

Table 4.27 Truth table for a Nibble-size minimum-delay complex binary multiplier [5, 8] (Minterm: $a_3a_2a_1a_0$ MUL $b_3b_2b_1b_0 = p_{11}p_{10}p_9p_8p_7p_6p_5p_4p_3p_2p_1p_0$)

Minterm	Multiplicand				Multiplier				Product
	a_3	a_2	a_1	a_0	b_3	b_2	b_1	b_0	$p_{11}p_{10}\cdots p_1p_0$
0	0	0	0	0	0	0	0	0	000000000000
1	0	0	0	0	0	0	0	1	000000000000
2	0	0	0	0	0	0	1	0	000000000000
3	0	0	0	0	0	0	1	1	000000000000
4	0	0	0	0	0	1	0	0	000000000000
5	0	0	0	0	0	1	0	1	000000000000
6	0	0	0	0	0	1	1	0	000000000000
7	0	0	0	0	0	1	1	1	000000000000
8	0	0	0	0	1	0	0	0	000000000000
9	0	0	0	0	1	0	0	1	000000000000
10	0	0	0	0	1	0	1	0	000000000000
11	0	0	0	0	1	0	1	1	000000000000
12	0	0	0	0	1	1	0	0	000000000000
13	0	0	0	0	1	1	0	1	000000000000
14	0	0	0	0	1	1	1	0	000000000000
15	0	0	0	0	1	1	1	1	000000000000
16	0	0	0	1	0	0	0	0	000000000000
17	0	0	0	1	0	0	0	1	000000000001
18	0	0	0	1	0	0	1	0	000000000010
19	0	0	0	1	0	0	1	1	000000000011
20	0	0	0	1	0	1	0	0	000000000100
21	0	0	0	1	0	1	0	1	000000000101
22	0	0	0	1	0	1	1	0	000000111010
23	0	0	0	1	0	1	1	1	000000000111
24	0	0	0	1	1	0	0	0	000000001000
25	0	0	0	1	1	0	0	1	000000001001
26	0	0	0	1	1	0	1	0	000000001010
27	0	0	0	1	1	0	1	1	000000001011
28	0	0	0	1	1	1	0	0	000000001100
29	0	0	0	1	1	1	0	1	000000001101
30	0	0	0	1	1	1	1	0	000000001110
31	0	0	0	1	1	1	1	1	000000001111

Table 4.28 Truth table for a Nibble-size minimum-delay complex binary multiplier [5, 8] (Minterm: $a_3a_2a_1a_0$ MUL $b_3b_2b_1b_0 = p_{11}p_{10}p_9p_8p_7p_6p_5p_4p_3p_2p_1p_0$)

Minterm	Multiplicand				Multiplier				Product
	a_3	a_2	a_1	a_0	b_3	b_2	b_1	b_0	$p_{11}p_{10}\cdots p_1p_0$
32	0	0	1	0	0	0	0	0	000000000000
33	0	0	1	0	0	0	0	1	000000000010
34	0	0	1	0	0	0	1	0	000000000100
35	0	0	1	0	0	0	1	1	000000000110
36	0	0	1	0	0	1	0	0	000000001000
37	0	0	1	0	0	1	0	1	000000001010
38	0	0	1	0	0	1	1	0	000000001100
39	0	0	1	0	0	1	1	1	000000001110
40	0	0	1	0	1	0	0	0	000000010000
41	0	0	1	0	1	0	0	1	000000010010
42	0	0	1	0	1	0	1	0	000000010100
43	0	0	1	0	1	0	1	1	000000010110
44	0	0	1	0	1	1	0	0	000000011000
45	0	0	1	0	1	1	0	1	000000011010
46	0	0	1	0	1	1	1	0	000000011100
47	0	0	1	0	1	1	1	1	000000011110
48	0	0	1	1	0	0	0	0	000000000000
49	0	0	1	1	0	0	0	1	000000000011
50	0	0	1	1	0	0	1	0	000000000110
51	0	0	1	1	0	0	1	1	000000011101
52	0	0	1	1	0	1	0	0	000000001100
53	0	0	1	1	0	1	0	1	000000001111
54	0	0	1	1	0	1	1	0	000000111010
55	0	0	1	1	0	1	1	1	000000000001
56	0	0	1	1	1	0	0	0	000000011000
57	0	0	1	1	1	0	0	1	000000011011
58	0	0	1	1	1	0	1	0	000000011110
59	0	0	1	1	1	0	1	1	001110100101
60	0	0	1	1	1	1	0	0	000001110100
61	0	0	1	1	1	1	0	1	000001110111
62	0	0	1	1	1	1	1	0	000000000010
63	0	0	1	1	1	1	1	1	000000011001

Table 4.29 Truth table for a Nibble-size minimum-delay complex binary multiplier [5, 8] (Minterm: $a_3a_2a_1a_0$ MUL $b_3b_2b_1b_0 = p_{11}p_{10}p_9p_8p_7p_6p_5p_4p_3p_2p_1p_0$)

Minterm	Multiplicand				Multiplier				Product
	a_3	a_2	a_1	a_0	b_3	b_2	b_1	b_0	$p_{11}p_{10}\cdots p_1p_0$
64	0	1	0	0	0	0	0	0	000000000000
65	0	1	0	0	0	0	0	1	000000000100
66	0	1	0	0	0	0	1	0	000000001000
67	0	1	0	0	0	0	1	1	000000001100
68	0	1	0	0	0	1	0	0	000000010000
69	0	1	0	0	0	1	0	1	000000010100
70	0	1	0	0	0	1	1	0	000000011000
71	0	1	0	0	0	1	1	1	000000011100
72	0	1	0	0	1	0	0	0	000000100000
73	0	1	0	0	1	0	0	1	000000100100
74	0	1	0	0	1	0	1	0	000000101000
75	0	1	0	0	1	0	1	1	000000101100
76	0	1	0	0	1	1	0	0	000000110000
77	0	1	0	0	1	1	0	1	000000110100
78	0	1	0	0	1	1	1	0	000000111000
79	0	1	0	0	1	1	1	1	000000111100
80	0	1	0	1	0	0	0	0	000000000000
81	0	1	0	1	0	0	0	1	000000000101
82	0	1	0	1	0	0	1	0	000000001010
83	0	1	0	1	0	0	1	1	000000001111
84	0	1	0	1	0	1	0	0	000000010100
85	0	1	0	1	0	1	0	1	000011100001
86	0	1	0	1	0	1	1	0	000000011110
87	0	1	0	1	0	1	1	1	000011101011
88	0	1	0	1	1	0	0	0	000000101000
89	0	1	0	1	1	0	0	1	000000101101
90	0	1	0	1	1	0	1	0	000111000010
91	0	1	0	1	1	0	1	1	000111000111
92	0	1	0	1	1	1	0	0	000000111100
93	0	1	0	1	1	1	0	1	011101001001
94	0	1	0	1	1	1	1	0	000111010110
95	0	1	0	1	1	1	1	1	000000100011

Table 4.30 Truth table for a Nibble-size minimum-delay complex binary multiplier [5, 8] (Minterm: $a_3a_2a_1a_0$ MUL $b_3b_2b_1b_0 = p_{11}p_{10}p_9p_8p_7p_6p_5p_4p_3p_2p_1p_0$)

Minterm	Multiplicand				Multiplier				Product
	a_3	a_2	a_1	a_0	b_3	b_2	b_1	b_0	$p_{11}p_{10}\cdots p_1p_0$
96	0	1	1	0	0	0	0	0	000000000000
97	0	1	1	0	0	0	0	1	000000000110
98	0	1	1	0	0	0	1	0	000000001100
99	0	1	1	0	0	0	1	1	000000111010
100	0	1	1	0	0	1	0	0	000000011000
101	0	1	1	0	0	1	0	1	000000011110
102	0	1	1	0	0	1	1	0	000001110100
103	0	1	1	0	0	1	1	1	000000000010
104	0	1	1	0	1	0	0	0	000000110000
105	0	1	1	0	1	0	0	1	000000110110
106	0	1	1	0	1	0	1	0	000000111100
107	0	1	1	0	1	0	1	1	011101001010
108	0	1	1	0	1	1	0	0	000011101000
109	0	1	1	0	1	1	0	1	000011101110
110	0	1	1	0	1	1	1	0	000000000100
111	0	1	1	0	1	1	1	1	000000110010
112	0	1	1	1	0	0	0	0	000000000000
113	0	1	1	1	0	0	0	1	000000000111
114	0	1	1	1	0	0	1	0	000000001110
115	0	1	1	1	0	0	1	1	000000000001
116	0	1	1	1	0	1	0	0	000000011100
117	0	1	1	1	0	1	0	1	000011101011
118	0	1	1	1	0	1	1	0	000000000010
119	0	1	1	1	0	1	1	1	000000011101
120	0	1	1	1	1	0	0	0	000000111000
121	0	1	1	1	1	0	0	1	000000111111
122	0	1	1	1	1	0	1	0	000111010110
123	0	1	1	1	1	0	1	1	000000111001
124	0	1	1	1	1	1	0	0	000000000100
125	0	1	1	1	1	1	0	1	000000110011
126	0	1	1	1	1	1	1	0	000000111010
127	0	1	1	1	1	1	1	1	000000000101

Table 4.31 Truth table for a Nibble-size minimum-delay complex binary multiplier [5, 8] (Minterm: $a_3a_2a_1a_0$ MUL $b_3b_2b_1b_0 = p_{11}p_{10}p_9p_8p_7p_6p_5p_4p_3p_2p_1p_0$)

Minterm	Multiplicand				Multiplier				Product
	a_3	a_2	a_1	a_0	b_3	b_2	b_1	b_0	$p_{11}p_{10}\cdots p_1p_0$
128	1	0	0	0	0	0	0	0	000000000000
129	1	0	0	0	0	0	0	1	000000001000
130	1	0	0	0	0	0	1	0	000000010000
131	1	0	0	0	0	0	1	1	000000011000
132	1	0	0	0	0	1	0	0	000000100000
133	1	0	0	0	0	1	0	1	000000101000
134	1	0	0	0	0	1	1	0	000000110000
135	1	0	0	0	0	1	1	1	000000111000
136	1	0	0	0	1	0	0	0	000001000000
137	1	0	0	0	1	0	0	1	000001001000
138	1	0	0	0	1	0	1	0	000001010000
139	1	0	0	0	1	0	1	1	000001011000
140	1	0	0	0	1	1	0	0	000001100000
141	1	0	0	0	1	1	0	1	000001101000
142	1	0	0	0	1	1	1	0	000001110000
143	1	0	0	0	1	1	1	1	000001111000
144	1	0	0	1	0	0	0	0	000000000000
145	1	0	0	1	0	0	0	1	000000001001
146	1	0	0	1	0	0	1	0	000000010010
147	1	0	0	1	0	0	1	1	000000011011
148	1	0	0	1	0	1	0	0	000000100100
149	1	0	0	1	0	1	0	1	000000101101
150	1	0	0	1	0	1	1	0	000000110110
151	1	0	0	1	0	1	1	1	000000111111
152	1	0	0	1	1	0	0	0	000001001000
153	1	0	0	1	1	0	0	1	001100100001
154	1	0	0	1	1	0	1	0	000001011010
155	1	0	0	1	1	0	1	1	001100110011
156	1	0	0	1	1	1	0	0	000001101100
157	1	0	0	1	1	1	0	1	111010000101
158	1	0	0	1	1	1	1	0	000001111110
159	1	0	0	1	1	1	1	1	111010010111

Table 4.32 Truth table for a Nibble-size minimum-delay complex binary multiplier [5, 8] (Minterm: $a_3a_2a_1a_0$ MUL $b_3b_2b_1b_0 = p_{11}p_{10}p_9p_8p_7p_6p_5p_4p_3p_2p_1p_0$)

Minterm	Multiplicand				Multiplier				Product
	a_3	a_2	a_1	a_0	b_3	b_2	b_1	b_0	$p_{11}p_{10}\cdots p_1p_0$
160	1	0	1	0	0	0	0	0	000000000000
161	1	0	1	0	0	0	0	1	000000001010
162	1	0	1	0	0	0	1	0	000000010100
163	1	0	1	0	0	0	1	1	000000011110
164	1	0	1	0	0	1	0	0	000000101000
165	1	0	1	0	0	1	0	1	000111000010
166	1	0	1	0	0	1	1	0	000000111100
167	1	0	1	0	0	1	1	1	000111010110
168	1	0	1	0	1	0	0	0	000001010000
169	1	0	1	0	1	0	0	1	000001011010
170	1	0	1	0	1	0	1	0	001110000100
171	1	0	1	0	1	0	1	1	001110001110
172	1	0	1	0	1	1	0	0	000001111000
173	1	0	1	0	1	1	0	1	111010010010
174	1	0	1	0	1	1	1	0	001110101100
175	1	0	1	0	1	1	1	1	000001000110
176	1	0	1	1	0	0	0	0	000000000000
177	1	0	1	1	0	0	0	1	000000001011
178	1	0	1	1	0	0	1	0	000000010110
179	1	0	1	1	0	0	1	1	001110100101
180	1	0	1	1	0	1	0	0	000000101100
181	1	0	1	1	0	1	0	1	000111000111
182	1	0	1	1	0	1	1	0	011101001010
183	1	0	1	1	0	1	1	1	000000111001
184	1	0	1	1	1	0	0	0	000001011000
185	1	0	1	1	1	0	0	1	001100110011
186	1	0	1	1	1	0	1	0	001110001110
187	1	0	1	1	1	0	1	1	001111111101
188	1	0	1	1	1	1	0	0	111010010100
189	1	0	1	1	1	1	0	1	111010011111
190	1	0	1	1	1	1	1	0	000001110010
191	1	0	1	1	1	1	1	1	000001000001

Table 4.33 Truth table for a Nibble-size minimum-delay complex binary multiplier [5, 8] (Minterm: $a_3a_2a_1a_0$ MUL $b_3b_2b_1b_0 = p_{11}p_{10}p_9p_8p_7p_6p_5p_4p_3p_2p_1p_0$)

Minterm	Multiplicand				Multiplier				Product
	a_3	a_2	a_1	a_0	b_3	b_2	b_1	b_0	$p_{11}p_{10}\cdots p_1p_0$
192	1	1	0	0	0	0	0	0	000000000000
193	1	1	0	0	0	0	0	1	000000001100
194	1	1	0	0	0	0	1	0	000000011000
195	1	1	0	0	0	0	1	1	000001110100
196	1	1	0	0	0	1	0	0	000000110000
197	1	1	0	0	0	1	0	1	000000111100
198	1	1	0	0	0	1	1	0	000011101000
199	1	1	0	0	0	1	1	1	000000000100
200	1	1	0	0	1	0	0	0	000001100000
201	1	1	0	0	1	0	0	1	000001101100
202	1	1	0	0	1	0	1	0	000001111000
203	1	1	0	0	1	0	1	1	111010010100
204	1	1	0	0	1	1	0	0	000111010000
205	1	1	0	0	1	1	0	1	000111011100
206	1	1	0	0	1	1	1	0	000000001000
207	1	1	0	0	1	1	1	1	000001100100
208	1	1	0	1	0	0	0	0	000000000000
209	1	1	0	1	0	0	0	1	000000001101
210	1	1	0	1	0	0	1	0	000000011010
211	1	1	0	1	0	0	1	1	000001110111
212	1	1	0	1	0	1	0	0	000000110100
213	1	1	0	1	0	1	0	1	011101001001
214	1	1	0	1	0	1	1	0	000011101110
215	1	1	0	1	0	1	1	1	000000110011
216	1	1	0	1	1	0	0	0	000001101000
217	1	1	0	1	1	0	0	1	111010000101
218	1	1	0	1	1	0	1	0	111010010010
219	1	1	0	1	1	0	1	1	111010011111
220	1	1	0	1	1	1	0	0	000111011100
221	1	1	0	1	1	1	0	1	000111000001
222	1	1	0	1	1	1	1	0	000001100110
223	1	1	0	1	1	1	1	1	000111011011

Table 4.34 Truth table for a Nibble-size minimum-delay complex binary multiplier [5, 8] (Minterm: $a_3a_2a_1a_0$ MUL $b_3b_2b_1b_0 = p_{11}p_{10}p_9p_8p_7p_6p_5p_4p_3p_2p_1p_0$)

Minterm	Multiplicand				Multiplier				Product
	a_3	a_2	a_1	a_0	b_3	b_2	b_1	b_0	$p_{11}p_{10}\cdots p_1p_0$
224	1	1	1	0	0	0	0	0	000000000000
225	1	1	1	0	0	0	0	1	000000001110
226	1	1	1	0	0	0	1	0	000000011100
227	1	1	1	0	0	0	1	1	000000000010
228	1	1	1	0	0	1	0	0	000000111000
229	1	1	1	0	0	1	0	1	000111010110
230	1	1	1	0	0	1	1	0	000000000100
231	1	1	1	0	0	1	1	1	000000111010
232	1	1	1	0	1	0	0	0	000001110000
233	1	1	1	0	1	0	0	1	000001111110
234	1	1	1	0	1	0	1	0	001110101100
235	1	1	1	0	1	0	1	1	000001110010
236	1	1	1	0	1	1	0	0	000000001000
237	1	1	1	0	1	1	0	1	000001100110
238	1	1	1	0	1	1	1	0	000001110100
239	1	1	1	0	1	1	1	1	000000001010
240	1	1	1	1	0	0	0	0	000000000000
241	1	1	1	1	0	0	0	1	000000001111
242	1	1	1	1	0	0	1	0	000000011110
243	1	1	1	1	0	0	1	1	000000011001
244	1	1	1	1	0	1	0	0	000000111100
245	1	1	1	1	0	1	0	1	000000100011
246	1	1	1	1	0	1	1	0	000000110010
247	1	1	1	1	0	1	1	1	000000000101
248	1	1	1	1	1	0	0	0	000001111000
249	1	1	1	1	1	0	0	1	111010010111
250	1	1	1	1	1	0	1	0	000001000110
251	1	1	1	1	1	0	1	1	000001000001
252	1	1	1	1	1	1	0	0	000001100100
253	1	1	1	1	1	1	0	1	000111011011
254	1	1	1	1	1	1	1	0	000000001010
255	1	1	1	1	1	1	1	1	000001111101

Table 4.35 Minterms corresponding to outputs of a Nibble-size minimum-delay multiplier [5, 8]

Multiplier Outputs	Corresponding minterms
p_{11}	157, 159, 173, 188, 189, 203, 217, 218, 219, 249
p_{10}	93, 107, 157, 159, 173, 182, 188, 189, 203, 213, 217, 218, 219, 249
p_9	59, 93, 107, 153, 155, 157, 159, 170, 171, 173, 174, 179, 182, 185, 186, 187, 188, 189, 203, 213, 217, 218, 219, 234, 249
p_8	59, 90, 91, 93, 94, 107, 122, 153, 155, 165, 167, 170, 171, 174, 179, 181, 182, 185, 186, 187, 204, 205, 213, 220, 221, 223, 229, 234, 253
p_7	59, 85, 87, 90, 91, 94, 108, 109, 117, 122, 157, 159, 165, 167, 170, 171, 173, 174, 179, 181, 186, 187, 188, 189, 198, 203, 204, 205, 214, 217, 218, 219, 220, 221, 223, 229, 234, 249, 253
p_6	60, 61, 85, 87, 90, 91, 93, 94, 102, 107, 108, 109, 117, 123, 136, 137, 138, 139, 140, 141, 142, 143, 152, 154, 156, 158, 165, 167, 168, 169, 172, 175, 181, 182, 184, 187, 190, 191, 195, 198, 200, 201, 202, 204, 205, 207, 211, 213, 214, 216, 220, 221, 222, 223, 229, 232, 233, 235, 237, 238, 248, 250, 251, 252, 253, 255
p_5	22, 54, 59, 60, 61, 72, 73, 74, 75, 76, 77, 78, 79, 85, 87, 88, 89, 92, 95, 99, 102, 104, 105, 106, 108, 109, 111, 117, 120, 121, 123, 125, 126, 132, 133, 134, 135, 140, 141, 142, 143, 148, 149, 150, 151, 153, 155, 156, 158, 164, 166, 172, 174, 179, 180, 183, 185, 187, 190, 195, 196, 197, 198, 200, 201, 202, 207, 211, 212, 214, 215, 216, 222, 228, 231, 232, 233, 234, 235, 237, 238, 244, 245, 246, 248, 252, 255
p_4	22, 40, 41, 42, 43, 44, 45, 46, 47, 51, 54, 56, 57, 58, 60, 61, 63, 68, 69, 70, 71, 76, 77, 78, 79, 84, 86, 92, 94, 99, 101, 102, 104, 105, 106, 111, 116, 119, 120, 121, 122, 123, 125, 126, 130, 131, 134, 135, 138, 139, 142, 143, 146, 147, 150, 151, 154, 155, 158, 159, 162, 163, 166, 167, 168, 169, 172, 173, 178, 183, 184, 185, 187, 188, 189, 190, 194, 195, 196, 197, 202, 203, 204, 205, 210, 211, 212, 215, 218, 219, 220, 223, 226, 228, 229, 231, 232, 233, 235, 238, 242, 243, 244, 246, 248, 249, 253, 255
p_3	22, 24, 25, 26, 27, 28, 29, 30, 31, 36, 37, 38, 39, 44, 45, 46, 47, 51, 52, 53, 54, 56, 57, 58, 63, 66, 67, 70, 71, 74, 75, 78, 79, 82, 83, 86, 87, 88, 89, 92, 93, 98, 99, 100, 101, 106, 107, 108, 109, 114, 116, 117, 119, 120, 121, 123, 126, 129, 131, 133, 135, 137, 139, 141, 143, 145, 147, 149, 151, 152, 154, 156, 158, 161, 163, 164, 166, 169, 171, 172, 174, 177, 180, 182, 183, 184, 186, 187, 189, 193, 194, 197, 198, 201, 202, 205, 206, 209, 210, 213, 214, 216, 219, 220, 223, 225, 226, 228, 231, 233, 234, 236, 239, 241, 242, 243, 244, 248, 253, 254, 255

Table 4.36 Minterms corresponding to outputs of a Nibble-size minimum-delay multiplier [5, 8]

Multiplier Outputs	Corresponding minterms
p_2	20, 21, 23, 28, 29, 30, 31, 34, 35, 38, 39, 42, 43, 46, 47, 50, 51, 52, 53, 58, 59, 60, 61, 65, 67, 69, 71, 73, 75, 77, 79, 81, 83, 84, 86, 89, 91, 92, 94, 97, 98, 101, 102, 105, 106, 109, 110, 113, 114, 116, 119, 121, 122, 124, 127, 148, 149, 150, 151, 156, 157, 158, 159, 162, 163, 166, 167, 170, 171, 174, 175, 178, 179, 180, 181, 186, 187, 188, 189, 193, 195, 197, 199, 201, 203, 205, 207, 209, 211, 212, 214, 217, 219, 220, 222, 225, 226, 229, 230, 233, 234, 237, 238, 241, 242, 244, 247, 249, 250, 252, 255
p_1	18, 19, 22, 23, 26, 27, 30, 31, 33, 35, 37, 39, 41, 43, 45, 47, 49, 50, 53, 54, 57, 58, 61, 62, 82, 83, 86, 87, 90, 91, 94, 95, 97, 99, 101, 103, 105, 107, 109, 111, 113, 114, 117, 118, 121, 122, 125, 126, 146, 147, 150, 151, 154, 155, 158, 159, 161, 163, 165, 167, 169, 171, 173, 175, 177, 178, 181, 182, 185, 186, 189, 190, 210, 211, 214, 215, 218, 219, 222, 223, 225, 227, 229, 231, 233, 235, 237, 239, 241, 242, 245, 246, 249, 250, 253, 254
p_0	17, 19, 21, 23, 25, 27, 29, 31, 49, 51, 53, 55, 57, 59, 61, 63, 81, 83, 85, 87, 89, 91, 93, 95, 113, 115, 117, 119, 121, 123, 125, 127, 145, 147, 149, 151, 153, 155, 157, 159, 177, 179, 181, 183, 185, 187, 189, 191, 209, 211, 213, 215, 217, 219, 221, 223, 241, 243, 245, 247, 249, 251, 253, 255

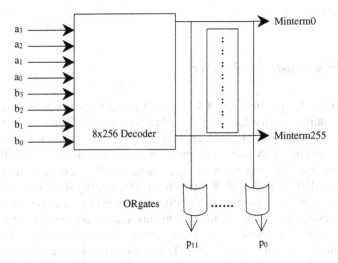

Fig. 4.7 Block diagram of a nibble-size minimum-delay complex binary multiplier [7]

4.3.2 Implementations

The minimum-delay multiplier has been implemented on various Xilinx FPGAs (Virtex V50CS144, Virtex2 2V10000FF1517, Spartan2 2S200PQ208, and Spartan2 2S30VQ100) and the statistics obtained are listed in Table 4.37 [7].

Table 4.37 Implementation statistics for nibble-size multiplier design on Xilinx FPGAs [7]

	Complex binary minimum-delay multiplier			
	Virtex V50CS144	Virtex2 2V10000FF1517	Spartan 2S200PQ208	Spartan 2S30PQ208
Number of external IOBs	20/94 (21 %)	20/1108 (1 %)	20/140 (14 %)	20/60 (33 %)
Number of slices	340/1200 (28 %)	340/61440 (1 %)	340/2352 (14 %)	340/432 (78 %)
Number of 4 input LUTs	676/2400 (28 %)	676/122880 (1 %)	676/4,704 (14 %)	676/864 (78 %)
Number of bonded IOBs	20/94 (21 %)	20/1108 (1 %)	20/140 (14 %)	20/60 (33 %)
Gate count	4617	4617	4617	4617
Maximum net delay (ns)	13.016	9.804	10.654	6.268
Maximum combinational delay (ns)	32.351	25.263	28.657	25.085

4.4 Divider Circuit for Complex Binary Numbers

A minimum-delay nibble-size divider for complex binary numbers has been presented in [5, 8].

4.4.1 Minimum-Delay Divider

The minimum-delay nibble-size CBNS divider has been designed following a procedure similar to what has been described for the adder, subtractor, and multiplier circuits in Sects. 4.1.1, 4.2.1, and 4.3.1 respectively. The procedure involves writing a truth table with four-bit dividend or numerator ($a_3a_2a_1a_0$) and divisor or denominator ($b_3b_2b_1b_0$) operands as inputs (total of $2^8 = 256$ minterms) and ten outputs ($R_9R_8R_7R_6R_5R_4R_3R_2R_1R_0$) which are obtained by dividing each pair of dividend and divisor according to the division procedure described in Chap. 3. Since the result of division operation contains small fractions for some sets of operands, it has been rounded off by converging numbers less than 0.5 to 0 and numbers greater than or equal to 0.5 to 1. This step is taken to avoid representing fractional numbers in CBNS. As a result, some degree of error in the final result has been introduced and, therefore, the resulting circuit can best be described as an "approximate" divider circuit. Each output of the divider circuit has been expressed in sum-of-minterms form and the design is implemented using an 8 × 256 decoder (to generate each minterm) and multiple-input OR gates (to combine relevant minterms for each output).

Tables 4.38, 4.39, 4.40, 4.41, 4.42, 4.43, 4.44, 4.45 present complete truth table for a nibble-size minimum-delay complex binary divider circuit and the sum-of-minterms expressions for outputs of the multiplier circuit are listed in Table 4.46. Block diagram of the divider circuit is given in Fig. 4.8.

Table 4.38 Truth table for a Nibble-size minimum-delay complex binary divider [5, 8] (Minterm: $a_3a_2a_1a_0$ DIV $b_3b_2b_1b_0 = R_9R_8R_7R_6R_5R_4R_3R_2R_1R_0$) NaN : Not a number

Minterm	Dividend				Divisor				Result
	a_3	a_2	a_1	a_0	b_3	b_2	b_1	b_0	$R_9R_8...R_1R_0$
0	0	0	0	0	0	0	0	0	NaN
1	0	0	0	0	0	0	0	1	0000000000
2	0	0	0	0	0	0	1	0	0000000000
3	0	0	0	0	0	0	1	1	0000000000
4	0	0	0	0	0	1	0	0	0000000000
5	0	0	0	0	0	1	0	1	0000000000
6	0	0	0	0	0	1	1	0	0000000000
7	0	0	0	0	0	1	1	1	0000000000
8	0	0	0	0	1	0	0	0	0000000000
9	0	0	0	0	1	0	0	1	0000000000
10	0	0	0	0	1	0	1	0	0000000000
11	0	0	0	0	1	0	1	1	0000000000
12	0	0	0	0	1	1	0	0	0000000000
13	0	0	0	0	1	1	0	1	0000000000
14	0	0	0	0	1	1	1	0	0000000000
15	0	0	0	0	1	1	1	1	0000000000
16	0	0	0	1	0	0	0	0	NaN
17	0	0	0	1	0	0	0	1	0000000001
18	0	0	0	1	0	0	1	0	0000000110
19	0	0	0	1	0	0	1	1	0000000111
20	0	0	0	1	0	1	0	0	0000000011
21	0	0	0	1	0	1	0	1	0000000000
22	0	0	0	1	0	1	1	0	0000000010
23	0	0	0	1	0	1	1	1	0000000011
24	0	0	0	1	1	0	0	0	0000000000
25	0	0	0	1	1	0	0	1	0000000000
26	0	0	0	1	1	0	1	0	0000000000
27	0	0	0	1	1	0	1	1	0000000000
28	0	0	0	1	1	1	0	0	0000000001
29	0	0	0	1	1	1	0	1	0000000000
30	0	0	0	1	1	1	1	0	0000111010
31	0	0	0	1	1	1	1	1	0000000000

Table 4.39 Truth table for a Nibble-size minimum-delay complex binary divider [5, 8]
(Minterm: $a_3a_2a_1a_0$ DIV $b_3b_2b_1b_0 = R_9R_8R_7R_6R_5R_4R_3R_2R_1R_0$)
NaN : Not a number

Minterm	Dividend				Divisor				Result
	a_3	a_2	a_1	a_0	b_3	b_2	b_1	b_0	$R_9R_8...R_1R_0$
32	0	0	1	0	0	0	0	0	NaN
33	0	0	1	0	0	0	0	1	0000000010
34	0	0	1	0	0	0	1	0	0000000001
35	0	0	1	0	0	0	1	1	0000001110
36	0	0	1	0	0	1	0	0	0000000110
37	0	0	1	0	0	1	0	1	0000011101
38	0	0	1	0	0	1	1	0	0000000111
39	0	0	1	0	0	1	1	1	0000000110
40	0	0	1	0	1	0	0	0	0000000011
41	0	0	1	0	1	0	0	1	0000000000
42	0	0	1	0	1	0	1	0	0000000000
43	0	0	1	0	1	0	1	1	0000000000
44	0	0	1	0	1	1	0	0	0000000010
45	0	0	1	0	1	1	0	1	0000000000
46	0	0	1	0	1	1	1	0	0000000011
47	0	0	1	0	1	1	1	1	0000000011
48	0	0	1	1	0	0	0	0	NaN
49	0	0	1	1	0	0	0	1	0000000011
50	0	0	1	1	0	0	1	0	0000111010
51	0	0	1	1	0	0	1	1	0000000001
52	0	0	1	1	0	1	0	0	0000011101
53	0	0	1	1	0	1	0	1	0000000000
54	0	0	1	1	0	1	1	0	0000000110
55	0	0	1	1	0	1	1	1	0000011101
56	0	0	1	1	1	0	0	0	0000000000
57	0	0	1	1	1	0	0	1	0000000000
58	0	0	1	1	1	0	1	0	0000000000
59	0	0	1	1	1	0	1	1	0000000000
60	0	0	1	1	1	1	0	0	0000000011
61	0	0	1	1	1	1	0	1	0000000000
62	0	0	1	1	1	1	1	0	0000001110
63	0	0	1	1	1	1	1	1	0000000000

Table 4.40 Truth table for a Nibble-size minimum-delay complex binary divider [5, 8] (Minterm: $a_3a_2a_1a_0$ DIV $b_3b_2b_1b_0 = R_9R_8R_7R_6R_5R_4R_3R_2R_1R_0$) NaN : Not a number

Minterm	Dividend				Divisor				Result
	a_3	a_2	a_1	a_0	b_3	b_2	b_1	b_0	$R_9R_8...R_1R_0$
64	0	1	0	0	0	0	0	0	NaN
65	0	1	0	0	0	0	0	1	0000000100
66	0	1	0	0	0	0	1	0	0000000010
67	0	1	0	0	0	0	1	1	0000011100
68	0	1	0	0	0	1	0	0	0000000001
69	0	1	0	0	0	1	0	1	0000000001
70	0	1	0	0	0	1	1	0	0000001110
71	0	1	0	0	0	1	1	1	0000001100
72	0	1	0	0	1	0	0	0	0000000110
73	0	1	0	0	1	0	0	1	0000000000
74	0	1	0	0	1	0	1	0	0000011101
75	0	1	0	0	1	0	1	1	0000000000
76	0	1	0	0	1	1	0	0	0000000111
77	0	1	0	0	1	1	0	1	0000000111
78	0	1	0	0	1	1	1	0	0000000110
79	0	1	0	0	1	1	1	1	0000000111
80	0	1	0	1	0	0	0	0	NaN
81	0	1	0	1	0	0	0	1	0000000101
82	0	1	0	1	0	0	1	0	0000011111
83	0	1	0	1	0	0	1	1	0011101011
84	0	1	0	1	0	1	0	0	0000001110
85	0	1	0	1	0	1	0	1	0000000001
86	0	1	0	1	0	1	1	0	0001110101
87	0	1	0	1	0	1	1	1	0000001111
88	0	1	0	1	1	0	0	0	0000000111
89	0	1	0	1	1	0	0	1	0000000111
90	0	1	0	1	1	0	1	0	0000000110
91	0	1	0	1	1	0	1	1	0000000111
92	0	1	0	1	1	1	0	0	0000111010
93	0	1	0	1	1	1	0	1	0000000111
94	0	1	0	1	1	1	1	0	0011101001
95	0	1	0	1	1	1	1	1	0000000111

Table 4.41 Truth table for a Nibble-size minimum-delay complex binary divider [5, 8] (Minterm: $a_3a_2a_1a_0$ DIV $b_3b_2b_1b_0 = R_9R_8R_7R_6R_5R_4R_3R_2R_1R_0$)
NaN : Not a number

Minterm	Dividend				Divisor				Result
	a_3	a_2	a_1	a_0	b_3	b_2	b_1	b_0	$R_9R_8...R_1R_0$
96	0	1	1	0	0	0	0	0	NaN
97	0	1	1	0	0	0	0	1	0000000110
98	0	1	1	0	0	0	1	0	0000000011
99	0	1	1	0	0	0	1	1	0000000010
100	0	1	1	0	0	1	0	0	0000111010
101	0	1	1	0	0	1	0	1	0000000111
102	0	1	1	0	0	1	1	0	0000000001
103	0	1	1	0	0	1	1	1	0000111010
104	0	1	1	0	1	0	0	0	0000011101
105	0	1	1	0	1	0	0	1	0000000000
106	0	1	1	0	1	0	1	0	0000000000
107	0	1	1	0	1	0	1	1	0000000000
108	0	1	1	0	1	1	0	0	0000000110
109	0	1	1	0	1	1	0	1	0000000000
110	0	1	1	0	1	1	1	0	0000011101
111	0	1	1	0	1	1	1	1	0000011101
112	0	1	1	1	0	0	0	0	NaN
113	0	1	1	1	0	0	0	1	0000000111
114	0	1	1	1	0	0	1	0	0000000010
115	0	1	1	1	0	0	1	1	0000011101
116	0	1	1	1	0	1	0	0	0000000001
117	0	1	1	1	0	1	0	1	0000000000
118	0	1	1	1	0	1	1	0	0000001110
119	0	1	1	1	0	1	1	1	0000000001
120	0	1	1	1	1	0	0	0	0000000000
121	0	1	1	1	1	0	0	1	0000000000
122	0	1	1	1	1	0	1	0	0000000000
123	0	1	1	1	1	0	1	1	0000000000
124	0	1	1	1	1	1	0	0	0000000111
125	0	1	1	1	1	1	0	1	0000000000
126	0	1	1	1	1	1	1	0	0000000110
127	0	1	1	1	1	1	1	1	0000000000

Table 4.42 Truth table for a nibble-size minimum-delay complex binary divider [5, 8] (Minterm: $a_3a_2a_1a_0$ DIV $b_3b_2b_1b_0 = R_9R_8R_7R_6R_5R_4R_3R_2R_1R_0$) NaN : Not a number

Minterm	Dividend				Divisor				Result
	a_3	a_2	a_1	a_0	b_3	b_2	b_1	b_0	$R_9R_8...R_1R_0$
128	1	0	0	0	0	0	0	0	NaN
129	1	0	0	0	0	0	0	1	0000001000
130	1	0	0	0	0	0	1	0	0000000100
131	1	0	0	0	0	0	1	1	0000111000
132	1	0	0	0	0	1	0	0	0000000010
133	1	0	0	0	0	1	0	1	0000000011
134	1	0	0	0	0	1	1	0	0000011100
135	1	0	0	0	0	1	1	1	0000011000
136	1	0	0	0	1	0	0	0	0000000001
137	1	0	0	0	1	0	0	1	0000000001
138	1	0	0	0	1	0	1	0	0000000001
139	1	0	0	0	1	0	1	1	0000000001
140	1	0	0	0	1	1	0	0	0000001110
141	1	0	0	0	1	1	0	1	0000001110
142	1	0	0	0	1	1	1	0	0000001100
143	1	0	0	0	1	1	1	1	0000000001
144	1	0	0	1	0	0	0	0	NaN
145	1	0	0	1	0	0	0	1	0000001001
146	1	0	0	1	0	0	1	0	0000110010
147	1	0	0	1	0	0	1	1	0000111111
148	1	0	0	1	0	1	0	0	0000011001
149	1	0	0	1	0	1	0	1	0001110100
150	1	0	0	1	0	1	1	0	0000011110
151	1	0	0	1	0	1	1	1	0000011011
152	1	0	0	1	1	0	0	0	0000000001
153	1	0	0	1	1	0	0	1	0000000001
154	1	0	0	1	1	0	1	0	0000000010
155	1	0	0	1	1	0	1	1	0000000001
156	1	0	0	1	1	1	0	0	0000001111
157	1	0	0	1	1	1	0	1	0000001110
158	1	0	0	1	1	1	1	0	0111010110
159	1	0	0	1	1	1	1	1	0000001100

Table 4.43 Truth table for a Nibble-size minimum-delay complex binary divider [5, 8] (Minterm: $a_3a_2a_1a_0$ DIV $b_3b_2b_1b_0 = R_9R_8R_7R_6R_5R_4R_3R_2R_1R_0$) NaN : Not a number

Minterm	Dividend				Divisor				Result
	a_3	a_2	a_1	a_0	b_3	b_2	b_1	b_0	$R_9R_8...R_1R_0$
160	1	0	1	0	0	0	0	0	NaN
161	1	0	1	0	0	0	0	1	0000001010
162	1	0	1	0	0	0	1	0	0000000101
163	1	0	1	0	0	0	1	1	0111010110
164	1	0	1	0	0	1	0	0	0000011111
165	1	0	1	0	0	1	0	1	0000000010
166	1	0	1	0	0	1	1	0	0011101011
167	1	0	1	0	0	1	1	1	0000011110
168	1	0	1	0	1	0	0	0	0000001110
169	1	0	1	0	1	0	0	1	0000001110
170	1	0	1	0	1	0	1	0	0000000001
171	1	0	1	0	1	0	1	1	0000000001
172	1	0	1	0	1	1	0	0	0001110101
173	1	0	1	0	1	1	0	1	0000000011
174	1	0	1	0	1	1	1	0	0000001111
175	1	0	1	0	1	1	1	1	0000001110
176	1	0	1	1	0	0	0	0	NaN
177	1	0	1	1	0	0	0	1	0000001011
178	1	0	1	1	0	0	1	0	0000111110
179	1	0	1	1	0	0	1	1	0000111001
180	1	0	1	1	0	1	0	0	0000011111
181	1	0	1	1	0	1	0	1	0000000010
182	1	0	1	1	0	1	1	0	0011101010
183	1	0	1	1	0	1	1	1	1110100101
184	1	0	1	1	1	0	0	0	0000000001
185	1	0	1	1	1	0	0	1	0000000001
186	1	0	1	1	1	0	1	0	0000000001
187	1	0	1	1	1	0	1	1	0000000001
188	1	0	1	1	1	1	0	0	0001110101
189	1	0	1	1	1	1	0	1	0000001110
190	1	0	1	1	1	1	1	0	0111010010
191	1	0	1	1	1	1	1	1	0000001110

Table 4.44 Truth table for a Nibble-size minimum-delay complex binary divider [5, 8] (Minterm: $a_3a_2a_1a_0$ DIV $b_3b_2b_1b_0 = R_9R_8R_7R_6R_5R_4R_3R_2R_1R_0$)
NaN : Not a number

Minterm	Dividend				Divisor				Result
	a_3	a_2	a_1	a_0	b_3	b_2	b_1	b_0	$R_9R_8...R_1R_0$
192	1	1	0	0	0	0	0	0	NaN
193	1	1	0	0	0	0	0	1	0000001100
194	1	1	0	0	0	0	1	0	0000000110
195	1	1	0	0	0	0	1	1	0000000100
196	1	1	0	0	0	1	0	0	0000000011
197	1	1	0	0	0	1	0	1	0000000011
198	1	1	0	0	0	1	1	0	0000000010
199	1	1	0	0	0	1	1	1	0001110100
200	1	1	0	0	1	0	0	0	0000111010
201	1	1	0	0	1	0	0	1	0000000000
202	1	1	0	0	1	0	1	0	0000000111
203	1	1	0	0	1	0	1	1	0000000000
204	1	1	0	0	1	1	0	0	0000000001
205	1	1	0	0	1	1	0	1	0000000001
206	1	1	0	0	1	1	1	0	0000111010
207	1	1	0	0	1	1	1	1	0000000001
208	1	1	0	1	0	0	0	0	NaN
209	1	1	0	1	0	0	0	1	0000001101
210	1	1	0	1	0	0	1	0	0011101000
211	1	1	0	1	0	0	1	1	0000110011
212	1	1	0	1	0	1	0	0	0001110100
213	1	1	0	1	0	1	0	1	0000001110
214	1	1	0	1	0	1	1	0	0000011000
215	1	1	0	1	0	1	1	1	0001110111
216	1	1	0	1	1	0	0	0	0000111010
217	1	1	0	1	1	0	0	1	0000000001
218	1	1	0	1	1	0	1	0	0000000111
219	1	1	0	1	1	0	1	1	0000000111
220	1	1	0	1	1	1	0	0	0000001100
221	1	1	0	1	1	1	0	1	0000000001
222	1	1	0	1	1	1	1	0	0000111000
223	1	1	0	1	1	1	1	1	0000111010

Table 4.45 Truth table for a Nibble-size minimum-delay complex binary divider [5, 8]
(Minterm: $a_3a_2a_1a_0$ DIV $b_3b_2b_1b_0 = R_9R_8R_7R_6R_5R_4R_3R_2R_1R_0$)
NaN : Not a number

Minterm	Dividend				Divisor				Result
	a_3	a_2	a_1	a_0	b_3	b_2	b_1	b_0	$R_9R_8...R_1R_0$
224	1	1	1	0	0	0	0	0	NaN
225	1	1	1	0	0	0	0	1	0000001110
226	1	1	1	0	0	0	1	0	0000000111
227	1	1	1	0	0	0	1	1	0000111010
228	1	1	1	0	0	1	0	0	0000000010
229	1	1	1	0	0	1	0	1	0000000011
230	1	1	1	0	0	1	1	0	0000011101
231	1	1	1	0	0	1	1	1	0000000010
232	1	1	1	0	1	0	0	0	0000000001
233	1	1	1	0	1	0	0	1	0000000000
234	1	1	1	0	1	0	1	0	0000000000
235	1	1	1	0	1	0	1	1	0000000000
236	1	1	1	0	1	1	0	0	0000001110
237	1	1	1	0	1	1	0	1	0000000000
238	1	1	1	0	1	1	1	0	0000000001
239	1	1	1	0	1	1	1	1	0000000001
240	1	1	1	1	0	0	0	0	NaN
241	1	1	1	1	0	0	0	1	0000001111
242	1	1	1	1	0	0	1	0	0011101001
243	1	1	1	1	0	0	1	1	0000000101
244	1	1	1	1	0	1	0	0	0000000010
245	1	1	1	1	0	1	0	1	0000000011
246	1	1	1	1	0	1	1	0	0000011111
247	1	1	1	1	0	1	1	1	0000011001
248	1	1	1	1	1	0	0	0	0000000001
249	1	1	1	1	1	0	0	1	0000000001
250	1	1	1	1	1	0	1	0	0000111010
251	1	1	1	1	1	0	1	1	0000000001
252	1	1	1	1	1	1	0	0	0000001110
253	1	1	1	1	1	1	0	1	0000000001
254	1	1	1	1	1	1	1	0	0000001111
255	1	1	1	1	1	1	1	1	0000000001

Table 4.46 Minterms corresponding to outputs of a Nibble-size minimum-delay divider [5, 8]

Divider Outputs	Corresponding Minterms
R_9	183
R_8	158, 163, 190
R_7	83, 94, 166, 182, 210, 242
R_6	86, 149, 172, 188, 199, 212, 215
R_5	30, 50, 92, 100, 103, 131, 146, 147, 178, 179, 200, 206, 211, 216, 222, 223, 227, 250
R_4	37, 52, 55, 67, 74, 82, 104, 110, 111, 115, 134, 135, 148, 150, 151, 164, 167, 180, 214, 230, 246, 247
R_3	35, 62, 70, 71, 84, 87, 118, 129, 140, 141, 142, 145, 156, 157, 159, 161, 168, 169, 174, 175, 177, 189, 191, 193, 209, 213, 220, 225, 236, 241, 252, 254
R_2	18, 19, 36, 38, 39, 54, 65, 72, 76, 77, 78, 79, 81, 88, 89, 90, 91, 93, 95, 97, 101, 108, 113, 124, 126, 130, 162, 194, 195, 202, 218, 219, 226, 243
R_1	20, 22, 23, 33, 40, 44, 46, 47, 49, 60, 66, 98, 99, 114, 132, 133, 154, 165, 173, 181, 196, 197, 198, 228, 229, 231, 244, 245
R_0	17, 28, 34, 51, 68, 69, 85, 102, 116, 119, 136, 137, 138, 139, 143, 152, 153, 155, 170, 171, 184, 185, 186, 187, 204, 205, 207, 217, 221, 232, 238, 239, 248, 249, 251, 253, 255

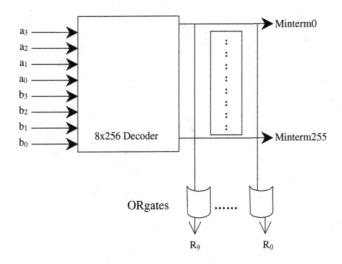

Fig. 4.8 Block diagram of a nibble-size minimum-delay complex binary divider

References

1. T. Jamil, B. Arafeh, A. AlHabsi, Design of nibble-size adder for (−1+j)-base complex binary numbers. Proc. World Multiconference Syst. Cybern. Inform. **5**, 297–302 (2002)
2. B. Arafeh, T. Jamil, A. AlHabsi, A nibble-size ripple-carry adder for (−1 + j)-base complex binary numbers. Proc. Int. Arab Conf. Inform. Technol. **1**, 207–211 (2002)
3. T. Jamil, B. Arafeh, A. AlHabsi, Hardware implementation and performance evaluation of complex binary adder designs. Proc. World Multiconference Syst. Cybernet. Inform. **2**, 68–73 (2003)
4. J. Goode, T. Jamil, D. Callahan, A simple circuit for adding complex numbers. WSEAS Trans. Informa. Sci. Appl. **1**(1), 61–66 (2004)
5. T. Jamil, Design of arithmetic circuits for complex binary number system. J. Am. Inst. Phys. IAENG Trans. Eng. Technol. **1373**(1), 83–97 (2011)
6. T. Jamil, A. Abdulghani, A. AlMaashari, Design of a nibble-size subtractor for (−1 + j)-base complex binary numbers. WSEAS Trans Circuits Syst **3**(5), 1067–1072 (2004)
7. T. Jamil, A. AlMaashari, A. Abdulghani, Design and implementation of a nibble-size multiplier for (−1 + j)-base complex binary numbers. WSEAS Trans Circuits Syst **4**(11), 1539–1544 (2005)
8. T. Jamil, S. AlAbri, Design of a divider circuit for complex binary numbers. Proc World Congr Eng Comp Sci **II**, 832–837 (2010)

Chapter 5
Complex Binary Associative Dataflow Processor Design

Abstract Complex Binary Number System provides an effective method to represent complex numbers in binary notation and, as discussed in the previous chapters, allows basic arithmetic operations to be performed on complex numbers with a better degree of efficiency. Associative dataflow paradigm provides a novel technique to do parallel processing within digital signal and image processing applications. It is, therefore, imperative to study the possibility of amalgamating the unique representation of complex numbers with an efficient parallel processing technique to come up with 'complex binary associative dataflow' processing. In this chapter, we are going to outline the design of a complex binary associative dataflow processor (CBADP) for which an Innovative Patent has been granted by the Australian Patent Office (IP-Australia).

5.1 Review of Complex Binary Number System

To completely understand the design of CBADP, let us first review complex binary number system [1]. The value of an n-bit complex binary number can be written in the form of a power series as follows:

$$a_{n-1}(-1+j)^{n-1} + a_{n-2}(-1+j)^{n-1}$$
$$+ a_{n-3}(-1+j)^{n-3} + a_{n-4}(-1+j)^{n-4}$$
$$+ \cdots$$
$$+ a_2(-1+j)^2 + a_1(-1+j)^1 + a_0(-1+j)^0 \quad (5.1)$$

where the coefficients $a_{n-1}, a_{n-2}, a_{n-3}, a_{n-4}, \ldots, a_2 a_1 a_0$ are binary (0 or 1) and $(-1+j)$ is the base of the CBNS. By applying the conversion algorithms described in Chap. 2, we can represent any given complex number in a unique single-unit binary string, as shown in the following examples:

T. Jamil, *Complex Binary Number System*,
SpringerBriefs in Electrical and Computer Engineering,
DOI: 10.1007/978-81-322-0854-9_5, © The Author(s) 2013

$$2012_{10} + j2012_{10} = 11101000000001110100011100000_{Base\,(-1+j)}$$

$$-60_{10} - j2000_{10} = 11101000000000011010110100000_{Base\,(-1+j)}$$

$$(0.351 + j0.351)_{Base\,10} = 0.0110100011110101111110001001\ldots_{Base(-1=j)}$$

$$(60.4375 + j60.4375)_{10} = 10000011101110.1000011_{Base\,(-1+j)}$$

The arithmetic operations in CBNS, discussed in Chap. 3, follow similar procedure as the traditional Base-2 number system with the exceptions that, in CBNS addition, $1_{10} + 1_{10} = 2_{10} = (1100)_{Base(-1+j)}$ and, in CBNS subtraction, $0_{10} - 1_{10} = -1_{10} = 11101_{Base(-1+j)}$. In CBNS multiplication, *zero rule* ($111 + 11 = 0$) plays an important role in reducing number of intermediate summands and, in CBNS division, we take the reciprocal of the denominator and multiply it with the numerator to get the result of division operation. Finally, in Chap. 4, we have presented individual designs of nibble-size adder, subtractor, multiplier, and divider circuits which can together be incorporated into an arithmetic and logic unit for complex binary numbers (CBALU).

5.2   What is Associative Dataflow Concept?

Of the currently prevalent ideas for building computers, the two well known and well developed are the control flow and the dataflow [2]. However, both these models are beset with limitations and weaknesses in exploiting parallelism to the utmost limit. Control-flow model lacks useful mathematical properties for program verification and is inherently sequential. The dataflow model, on the other hand, is based on partial ordering of the execution model and offers many attractive properties for parallel processing, including asynchrony and freedom from side-effects. However, a closer examination of the problems linked with dataflow model of computation reveals that they are mainly the by-products of using tokens during computations. These tokens need to be matched up with their partner token(s) prior to any operation to be carried out. This involves a time-consuming search process which results in degradation of overall performance. Since associative or content-addressable memories (AMs or CAMs) allow for a parallel search at a much faster rate, their use within the dataflow environment has been investigated under the concept of associative dataflow.

Eliminating the need for generating and handling tokens during the execution of a program, the associative dataflow model of computation processes a dataflow graph (program) in two phases: the search phase and the execution phase. During the search phase, the dataflow graph is conceptually assumed to be upside down and each node at the top of the hierarchy is considered to be the parent of the nodes which are connected to it through the arcs, referred to as children. Taking advantage

Fig. 5.1 Dataflow graph to
compute X = a + b +
c + d [1]

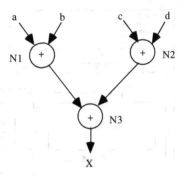

Fig. 5.2 Dataflow graph to
compute X = a + b +
c + d inverted to allow
progress of search phase [2]

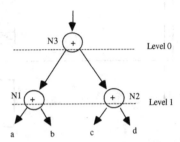

of the parallel search capabilities rendered by associative memories, the idea behind the search phase is for each parent node to search for its children. Once this search is completed, each node will know what its operands are and where the destination node(s) for the result is. During the execution phase, the operations will be performed as in conventional dataflow paradigm except the fact that now the matching of tokens will no longer be required. Thus, by eliminating tokens from the dataflow environment and using the search capabilities of associative memories, better performance can be achieved in parallel computers.

To better understand the concept of parent and children nodes, let us consider a simple dataflow graph to compute X = a + b + c + d (Fig. 5.1). The search phase of the associative dataflow concept requires that the given dataflow graph be turned upside-down in order for each parent to search for its children. The inverted dataflow graph to allow progress of this search phase is shown in Fig. 5.2, wherein the node at the top (N3) is at level 0, and the nodes N1 and N2 are at level 1. Node at level 0, i.e., N3, is the parent of the nodes at level 1, i.e., N1 and N2, or in other words, the nodes N1 and N2 at level 1 are the children of the node N3 at level 0. Similarly, operands' pairs (a,b) and (c,d) are the children of the nodes N1 and N2, respectively. During the search phase, each parent node will search for its children and, during the execution phase, the operations will be performed as in conventional dataflow paradigm, except the fact that now there will be no delay due to the matching of the tokens.

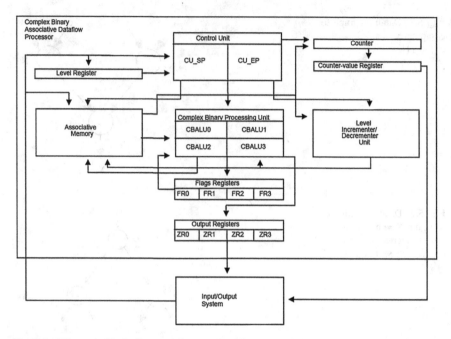

Fig. 5.3 Schematic block diagram of a complex binary associative dataflow processor [3]

5.3 Complex Binary Associative Dataflow Processor

Utilizing CBNS to represent complex numbers within associative dataflow processing (ADP) environment will enable us to take the best of both worlds (computer arithmetic and computer architecture) in an effort to achieve better degree of efficiency within digital signal and image processing applications. CBADP, which aims to combine CBNS with ADP, is the hardware realization of these efforts. A schematic block diagram for CBADP is shown in Fig. 5.3.

CBADP consists of following components:

(i)   Associative memory (AM) to collect and store the data needed for carrying out the given parallel computation, to store the dataflow graph in a format so as to permit the implementation of search phase of the associative dataflow concept, and to feed the data to the Processing Unit for computation of the result.

(ii) Processing unit (PU) containing four complex binary arithmetic and logic units (CBALUs) to compute the results of the operations carried out on the operands (represented in CBNS) and to set appropriate flags in the flags registers (FRs), to forward these results to the appropriate word within the AM (for onward processing at the next dataflow graph level) or to the output registers (ZRs) (for final result).

(iii) Level incrementer/decrementer unit (LIDU) to increment the current level number by one and to forward the new level number to the AM during the search phase or to decrement the current level number by one and to forward the new level number to the PU during the execution phase.

(iv) Control unit (CU) is hardwired with the task of generating appropriate control signals for search phase and execution phase.

(v) Counter-value register (CR) is used to store the counter value at the completion of each successful search phase.

(vi) Level register (LR) contains information about the maximum level number in the given dataflow graph.

(vii) Flags registers (FRs), one register corresponding to each CBALU, store the flags as a result of completion of an operation.

(viii) Output registers (ZRs), one register corresponding to each CBALU, store the result of the operation and make it available to the input/output system (IOS) for reading purposes.

5.4 Australian Innovation Patent No. 2010100706

An Australian Innovation Patent No. 2010100706 entitled Complex Binary Associative Dataflow Processor, describing the above design, has been granted in July 2010, details of which can be accessed at the following website address: http://pericles.ipaustralia.gov.au/ols/auspat/applicationDetails.do?applicationNo= 2010100706.

References

1. T. Jamil, in *Design of a complex binary associative dataflow processor. Proceedings of the 4th International Conference on Computer Engineering and Technology*, pp. 32–35 (2012)
2. T. Jamil, Introduction to associative dataflow processing—from concept to implementation. (VDM Verlag, Germany, 2010), ISBN-13: 9783639252330
3. T. Jamil, Complex binary associative dataflow processor (2010), http://pericles.ipaustralia. gov.au/ols/auspat/applicationDetails.do?applicationNo=2010100706

Chapter 6
Conclusion and Further Research

Abstract Complex Binary Number System (CBNS) with its uniqueness in representing complex numbers as a one-unit binary string holds great potential in the computer systems of tomorrow. With an innovative technique for parallel processing, such as associative dataflow, utilizing CBNS for representation of complex numbers, it is possible to leapfrog the speed of computing within today's signal and image processing applications. Preliminary work, spanning over two decades of research work and presented in this book, has shown good potential in this arena and scientists and engineers are urged to explore this avenue in the years to come. Although simulations of CBADP within digital signal and image processing applications and estimating performance evaluations will be very useful in the theoretical areas of computer architecture research, a complete working implementation of CBADP on a FPGA or an ASIC should be the ultimate goal of any researcher in this area.

T. Jamil, *Complex Binary Number System*,
SpringerBriefs in Electrical and Computer Engineering,
DOI: 10.1007/978-81-322-0854-9_6, © The Author(s) 2013